WAVES

DIMENSIONS OF MATHEMATICS

Waves

Alan J. Davies

The University of Hertfordshire

MACMILLAN

First published 1993 by
THE MACMILLAN PRESS LTD
Houndmills, Basingstoke, Hampshire RG21 2XS
and London
Companies and representatives
throughout the world

ISBN 0–333–54112–X

A catalogue record for this book is available
from the British Library

Printed in Hong Kong

The author and publishers wish to thank The Hulton-Deutsch Collection for the photo on page 6.

CONTENTS

PREFACE

This book provides an introduction to the phenomenon of wave motion. The material would usually be covered in an elementary course in applied mathematics at undergraduate level. However, it has been written so that it should be accessible to upper sixth form pupils who are studying mathematics and who wish to extend their interest further.

The only mathematical assumption that I have made is that the reader is familiar with the techniques of algebra and calculus usually found in 'A' level pure mathematics. Consequently, some parts of the text are developed in a manner which is mathematically less elegant than otherwise would be the case, e.g. I make no reference at all to partial differentiation in relation to the wave equation, nor to eigenvalues in relation to normal modes.

As far as the applied mathematics is concerned, I have assumed that the reader is familiar with Newton's second law. A knowledge of simple harmonic motion would be very useful, but has not been assumed.

Throughout the text, the reader is encouraged to develop some of the results via *PFTAs*: '*pauses for thought and action*'. These exercises form an important part of the learning process and as such involve the reader in an active rather than a passive role.

However, it is quite possible to read the text without attempting all of the exercises. Where appropriate, I have included simple experiments to illustrate more clearly the ideas being discussed. These can be performed at home using easily available equipment. Also, I have indicated where the reader with a personal computer could produce simple programs to illustrate some of the more complicated results. There are many programming languages now available on PCs and so I have avoided being specific by giving the program segments in pseudo-code. In the pseudo-code segments, the command set_polyline (n, x, y) is the command to draw the set of straight lines joining the n points whose Cartesian coordinates are given in the arrays x and y. The interested reader should have little difficulty producing programs in any suitable language.

Finally, I hope that the book provides a stimulation to find out more about waves. The bibliography provides a short list of textbooks and articles which could form a source for further study of wave motion.

Alan Davies

1 INTRODUCTION

1.1 WHAT ARE WAVES?

It is interesting to note that everybody who writes about waves begins by saying how difficult it is to produce a definition of what a wave is. We shall be no different in this introductory text.

Perhaps we could start by considering the 'everyday' idea of a wave. For most people, a wave would mean a wave on the sea or on a lake, and it would probably include some idea of the wave travelling along. Another possibility, which doesn't contain the idea of movement, might occur in answer to the question: What does a wave look like? An answer could well suggest something like the waves in someone's hair after a 'perm', or perhaps a wavy line. This particular pictorial view is so distinctive that a well-known chain of grocery stores uses it for its name. We could ask: How does waving someone goodbye tie in with our intuitive idea of a wave? There is motion involved but, apparently, not the same sort of motion as a wave rolling up a beach. Finally, many millions of people watched the incredible human wave in the famous Aztec Stadium in Mexico City during the 1986 World Cup Soccer Finals. There was no doubt in anybody's mind that this was a wave; certainly it was seen to be very similar to a wave on the sea.

So far so good. We now have an idea of what we wish to describe in more detail, so let's see if we can take things a little further. We would probably all agree that it is very easy to produce a wave: one way would be to shake the end of a piece of rope, a skipping rope or a washing line, for example; another way would be to drop a pebble into a pond. In both cases, we see ripples moving along, travelling away from the source of the disturbance. In the first case, the disturbance is propagated in one dimension, along the length of the string; in the second case, the disturbance is propagated in two dimensions – a sequence of circular ripples spreads out over the surface of the pond.

What are the fundamental points to be drawn from these two examples?

1

Firstly, for a wave to propagate, i.e. move along, it needs a medium in which it can travel. Secondly, it needs an initial disturbance in that medium. For waves on a string, the medium is the string itself while the disturbance is the displacement of the string. For waves on a pond, the medium is the surface of the water while the disturbance is the upward or downward displacement of the surface. Finally, in both cases, although the wave itself can propagate over quite some distance, the material comprising the medium has no substantial overall movement. For example, when water waves pass underneath a small boat floating on the surface, the boat merely bobs vertically up and down despite the fact that the waves themselves are clearly travelling in a horizontal direction.

Note here that even though no material is propagated, energy is transmitted by this water wave. Indeed, there have been a variety of research programmes aimed at developing techniques to extract energy from sea waves. At the time of writing, there are no large-scale installations. However, a typical experimental wave generator is the rocking-boom converter, shown in Figure 1.1.

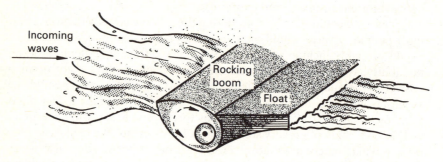

Figure 1.1 The rocking-boom converter. As the incoming wave strikes the boom, it transfers energy to the rocking motion of the boom.

We can illustrate the phenomenon of small overall movement of the medium with the aid of a simple 'wave machine'. This can be made using curtain tape and drinking straws.

Take a piece of curtain tape about 2 m long and insert a drinking straw through each pair of holes, leaving about 5 cm clear at each end of the tape. Fix one end of the tape and pull it tight at the other end. A wave can be propagated along the tape by giving the end a sharp twist. Notice that each straw remains at rest until the wave reaches it. When the wave arrives at each individual straw, that straw rocks backwards and then forwards, returning to its original position. Overall, each straw has moved no distance at all, but the wave has passed from one end of the tape to the other. A photograph of the 'wave machine' is shown in Figure 1.2.

As well as the two wave examples considered so far, sound waves are a

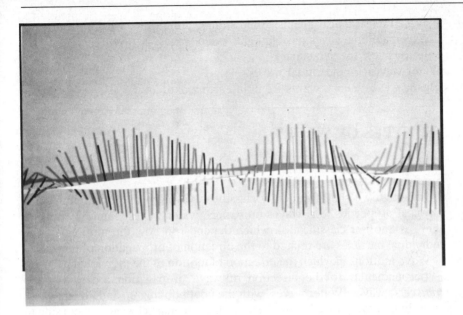

Figure 1.2 A 'wave machine' made from curtain tape and drinking straws.

familiar idea to most people, although it is not at all obvious that they are the same sort of phenomenon. However, they have yet another important property: they can transmit information. When someone speaks to us, sound waves are propagated through the air. When the many different vibrations reach our ear, the brain 'sorts out' these vibrations and interprets them as language.

To summarise, then, we can describe the important features of a wave as follows: a wave is a *disturbance* which is *propagated* in a *medium* with *little overall disturbance of the material*. The wave carries *energy* with it and has the ability also to *transmit information*. These features, together with a number of other properties, will be described, for waves in a variety of media, in the chapters which follow.

PFTA 1.1

Give some examples where wave motion is significant.

Solution 1.1

The following is a list but it is by no means exclusive:

3

Waves in musical instruments: violin strings, drum skins, wind instruments.
Electromagnetic waves: visible light, radio waves, X-rays.
Sound waves: speech, sonar.
Water waves: ripples, tidal waves.
Elastic waves: shock waves from an earthquake.

1.2 TYPES OF WAVES

The different types of wave, considered in Chapters 2 to 5, are described briefly in this section.

Although we shall frequently classify waves according to their specific physical properties, e.g. waves on strings, water waves, sound waves etc, there is another classification which depends on how the motions of the individual particles are related to the direction of propagation of the wave.

Wave motions in which the direction of motion of the individual particles is perpendicular to the direction of wave propagation are said to be *transverse* waves. Water waves, with the boat bobbing up and down, waves on a shaken *string* (see Figure 1.3), in which the particles move from side to side, and the Aztec wave, in which individual people stand up and then sit down again, are all examples of transverse waves.

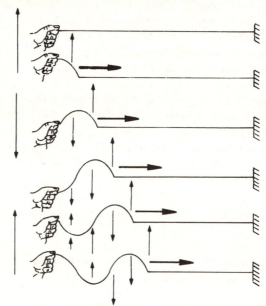

Figure 1.3 Propagation of a transverse wave caused by shaking one end of a string. A single shake will send out a single pulse while a steadily oscillating shake will send out a continuous wave.

Wave motions in which the direction of motion of the individual particles is back and forth along the direction of propagation of the wave are said to be *longitudinal* waves. Sound waves are an example of longitudinal waves. However, probably the simplest example to visualise is that of a wave in a *spring*.

Take a Slinkey spring and attach one end to the ground. Hold the other end and move it up and down so that it oscillates in a vertical direction. A sequence of pulses will propagate down the spring (see Figure 1.4). There are successive regions of *compression* and *extension* which form the propagated disturbances.

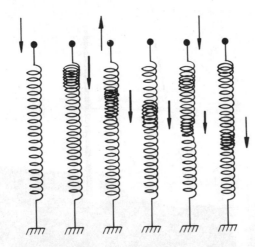

Figure 1.4   Longitudinal waves in a Slinkey spring.

PFTA 1.2

For each of the examples that you gave in PFTA 1.1, state whether the wave is transverse or longitudinal.

Solution 1.2

Transverse waves: violin strings, drum skins, electromagnetic waves, water waves.
Longitudinal waves: wind instruments, speech, sonar.
Shock waves from earthquakes may be either transverse, e.g. the so-called S wave, or longitudinal, e.g. the so-called P wave (see page 67).

Waves on strings occur in many places, e.g. vibrations of a piano or violin string, the vibrations of overhead wires for electric trains etc. *Waves in springs* occur, for example, in suspension systems. We shall discuss strings and springs in Chapter 4.

Longitudinal waves in *rods* are very similar to waves in springs. They are propagated in the same way due to the elastic properties of the material. Rods may also produce transverse and twisting, or torsional, vibrations. Torsional vibrations are rather complicated and will not be considered in detail. However, the form of such vibrations is illustrated very nicely in the photograph in Figure 1.5.

Figure 1.5 Torsional vibrations of the Tacoma Narrows bridge, USA.

Waves in membranes are very similar in nature to waves on strings. Obvious examples are waves on a drum skin. However, the sounds produced by a drum are not as 'musical' as those produced by, say, a violin. We shall see why that is in Chapter 5.

We have already mentioned sound waves. These are longitudinal vibrations which can be propagated in air due to its compressibility. We can think of this compressibility as the same sort of property as the elasticity in a spring. Sound waves can, in fact, be propagated in any gas or liquid; it is this property that is used in underwater sonar equipment. Sound waves will be considered in detail in Chapter 5.

We started by discussing one-dimensional waves. We then introduced

Figure 1.6 The megaphone guides sound waves in a particular direction.

two-dimensional waves, i.e. waves on the surface of a pond. Now, we consider three-dimensional waves.

Suppose that a person stands on a step-ladder and presses a fog horn, so that a sound pulse is emitted equally strongly in all directions. This pulse will propagate through the air as a spherical shell. The intensity of the sound will clearly decrease as the radius of the shell increases, since we can imagine that the energy is being spread over an ever-increasing surface area. Eventually, the sound will hardly be audible. We can improve matters by directing the sound using a megaphone (see Figure 1.6). Such a device is an example of a *waveguide*.

PFTA 1.3

Give some examples in which waves are directed by means of a waveguide.

Solution 1.3

Similar to the megaphone is the body of a musical wind instrument.

7

Another example, which is of increasing application, is the optical fibre, which acts as a guide for light waves.

We have seen, then, that waves are propagated by virtue of the vibration of individual particles. If we are to understand the properties of wave motion, it is essential that we understand, in detail, the motion of a single particle, and how that motion is affected when it is connected to one or more other particles. We consider these problems in Chapter 2, and then show how they tie in with wave motion in Chapter 3.

There are many general properties of waves which may be possessed by more than one type of wave, e.g. Doppler effect and interference, and we shall introduce these ideas whenever appropriate in the text. A brief discussion of some of the general properties of waves is given in Chapter 6.

Finally, there are two important types of wave which are a little too complicated to consider in this introductory text: water waves and electromagnetic phenomena.

Water waves are most usually caused by the action of wind on the water surface. In general, the higher the windspeed, the greater is the height of the wave. However, some of the most spectacular water waves are caused by other natural occurrences, such as earthquakes, volcanoes etc. Water waves are usually classified as either *tidal waves*, i.e. waves in shallow water, or *surface waves*, i.e. waves in deep water.

Electromagnetic phenomena may be propagated as waves, e.g. visible light, radio waves etc. However, there is a major difference here in that there is no motion of the medium. Furthermore, an equally useful model is that of light behaving like a particle. However, this so-called wave-particle duality requires mathematics way beyond the scope of this book.

2 VIBRATIONS

2.1 VIBRATIONS OF A SINGLE PARTICLE

We have seen, in the introduction in Chapter 1, that wave motion is associated with the oscillations of the individual particles which comprise the medium in which the wave propagates. The medium does, of course, contain many millions of particles and it would be a very ambitious task to try to describe a wave by considering the motion of every individual particle. However, there is much to be gained by becoming familiar with the properties of vibrations of a single particle. We shall consider the case of a particle of mass m attached to one end of a spring of stiffness k, the other end of which is fixed.

Before we start the analysis, let us see what is meant by the stiffness of a spring. The concept may be illustrated by the following simple experiment which can be done at home using an elastic string formed by joining together a number of elastic bands. You can, of course, use a suitable spring.

PFTA 2.1

Fix one end of the string to, say, the top of a door frame, as shown in Figure 2.1, and attach a large, empty yoghurt pot at the other end. Gradually place marbles, or any other suitable objects, in the pot, measuring the extension each time. Plot a graph of extension against number of marbles.

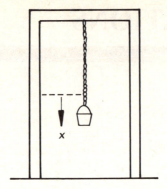

Figure 2.1 Home experiment with a simple elastic string.

Solution 2.1

The tension in the string is proportional to the number of marbles in the pot. In Figure 2.2, values of x, the extension, have been plotted against n, the number of marbles in the pot. The actual numerical values will depend on the type of elastic band used. However, Figure 2.2 shows a typical graph.

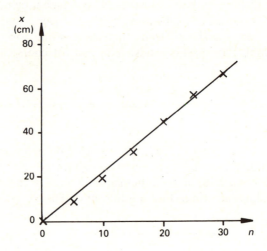

Figure 2.2 A plot of extension, x, against the number of marbles in the pot, n.

The graph in Figure 2.2 is approximately a straight line, so the relationship between the tension, F, in the string and the extension, x, can be written in the form:

$$F = kx \qquad\qquad (2.1)$$

where k is called the *stiffness* of the string. Equation (2.1) is known as *Hooke's law* for the string.

In reality, the elastic string differs from a spring in a very important way: the string can only sustain tension, it cannot sustain compression, whereas a spring can sustain both. Nevertheless, the simple experiment is satisfactory for our purposes.

We can also use the elastic string to introduce vibrations. Place some marbles in the pot, draw the pot below its static position and then release it. Notice that the system performs oscillations and that the time taken for each oscillation is approximately the same.

This type of motion is called *simple harmonic motion (SHM)*; it is a very important type of motion. An experiment to illustrate some of the ideas of SHM is described in PFTA 2.2 and PFTA 2.3. Working through these two exercises will help you to understand the concept of SHM and will introduce some of the terminology. The string in PFTAs 2.2 and 2.3 is the string you used in PFTA 2.1.

PFTA 2.2

Measure the length of the unstretched string. This is known as its *natural length*. Place some marbles in the pot and make a note of its position. The string will be stretched and will remain at rest at the *equilibrium position*. Pull the pot down a little and release it. Notice that it oscillates about the equilibrium position. The time taken for the pot to perform one complete oscillation is known as the *period*. Measure this period by timing ten complete oscillations and taking the average value. You should notice that the maximum displacement above and below the equilibrium position is the same and is almost constant over the ten oscillations. This maximum displacement is called the *amplitude*. Repeat the experiment with different values of the amplitude. Provided that you don't pull the pot down too far, you should notice that the period remains the same, i.e. it is independent of the amplitude.

PFTA 2.3

Repeat the experiment to find the period, T, as described in PFTA 2.2, using different numbers, n, of marbles in the pot. Tabulate the values of T and n and plot a graph of T against n.

Solution 2.3

The following table contains values from a typical experiment:

n	0	5	10	15	20	25	30
T(s)	0	0.647	0.920	1.187	1.427	1.640	1.827

The graph of T against n is shown in Figure 2.3.

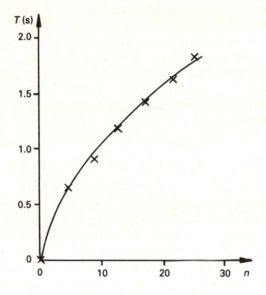

Figure 2.3 A plot of period, T, against n, the number of marbles.

The two main points that arise from these exercises are that the oscillations occur about an equilibrium position and that they have a period which is independent of the amplitude.

Newton's second law allows us to develop a mathematical model of SHM. We use it to write down the equation of motion of a spring–mass system. This system is one in which a single particle moves in one dimension. Such a system is said to have *one degree of freedom*.

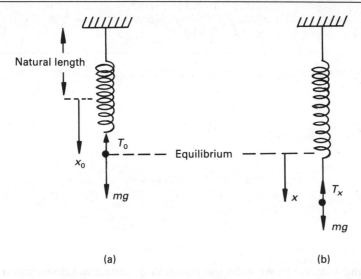

Figure 2.4 A spring–mass system in motion under the action of gravity: (a) equilibrium situation; (b) the system in motion.

Consider a particle, mass m, suspended in equilibrium at the end of a spring of stiffness k, as shown in Figure 2.4(a). We use the notation defined in Figure 2.4. In equilibrium, using Hooke's law:

$$T_0 = kx_0 \qquad (2.2)$$

However, in equilibrium, the tension, T_0, must support the weight, mg; that is:

$$T_0 = mg \qquad (2.3)$$

From equations (2.2) and (2.3):

$$kx_0 = mg$$
$$x_0 = \frac{mg}{k} \qquad (2.4)$$

Let us now consider a further extension, x, from the equilibrium position (see Figure 2.4(b)). Using Hooke's law:

$$T_x = k(x + x_0)$$

From equation (2.4):

$$T_x = k\left(x + \frac{mg}{k}\right) \tag{2.5}$$

Using Newton's second law for the particle in Figure 2.4(b):

$$mg - T_x = m\ddot{x}$$

and using equation (2.5):

$$mg - k\left(x + \frac{mg}{k}\right) = m\ddot{x}$$
$$m\ddot{x} = -kx$$
$$\ddot{x} = -\frac{kx}{m}$$

Replacing k/m by ω^2, we have the equation of simple harmonic motion:

$$\ddot{x} = -\omega^2 x \qquad \text{where } \omega^2 = \frac{k}{m} \tag{2.6}$$

This equation is often written in the equivalent form $\ddot{x} + \omega^2 x = 0$. The general solution of this equation is:

$$x = P \cos \omega t + Q \sin \omega t \tag{2.7}$$

In equation (2.7), P and Q are constants.

To develop this solution, we really need techniques for solving differential equations. However, for the purposes of this text, we shall simply show that x, given by equation (2.7), satisfies equation (2.6).

PFTA 2.4

By differentiating twice with respect to t, show that x, as given by equation (2.7), satisfies equation (2.6).

Solution 2.4

$$x = P \cos \omega t + Q \sin \omega t$$
$$\dot{x} = -\omega P \sin \omega t + \omega Q \cos \omega t$$

$$\ddot{x} = -\omega^2 P \cos \omega t - \omega^2 Q \sin \omega t$$
$$= -\omega^2 (P \cos \omega t + Q \sin \omega t)$$
$$= -\omega^2 x$$

PFTA 2.5

Show that the expression $P \cos \omega t + Q \sin \omega t$ can be written as:

$$A \cos (\omega t + \phi) \qquad \text{where } A = (P^2 + Q^2)^{1/2} \text{ and } \tan \phi = -\frac{Q}{P}$$

The general solution of equation (2.6) can now be written as:

$$x = A \cos (\omega t + \phi) \qquad \qquad \textbf{(2.8)}$$

Sketch the graph of x against t.

Solution 2.5

$$x = P \cos \omega t + Q \sin \omega t$$

$$= A \left(\frac{P}{A} \cos \omega t + \frac{Q}{A} \sin \omega t \right)$$

If $A = (P^2 + Q^2)^{1/2}$, then we may write $P/A = \cos \alpha$ and $Q/A = \sin \alpha$, as shown in Figure 2.5. Hence:

$$x = A(\cos \alpha \cos \omega t + \sin \alpha \sin \omega t)$$
$$= A \cos (\omega t - \alpha)$$

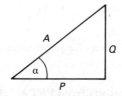

Figure 2.5 Definition of the angle α.

15

Now $\tan \alpha = Q/P$ so that if $\phi = -\alpha$, then $\tan \phi = -Q/P$ and:

$$x = A \cos (\omega t + \phi)$$

The graph of x against t is shown in Figure 2.6.

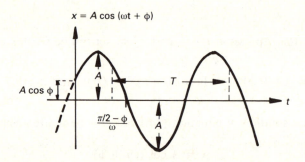

Figure 2.6 A sketch of the function $A \cos (\omega t + \phi)$.

The quantity A is called the *amplitude* of the vibration, ω is called the *angular frequency* and ϕ is called the *phase constant*. The *period*, T, is related to ω by:

$$T = \frac{2\pi}{\omega} \qquad (2.9)$$

In later chapters, we shall refer to the *frequency*, f, where $f = 1/T$. It is important not to confuse frequency with angular frequency. Unfortunately, it is often the case that the term frequency is used when what is really meant is angular frequency. The parameters A, ω, ϕ and T of the motion are illustrated in Figure 2.6.

Returning to the spring–mass system for a moment, notice that, since $\omega^2 = k/m$, it follows that:

$$T = 2\pi \sqrt{\frac{m}{k}}$$

PFTA 2.6

Using your experimental results from PFTA 2.3, plot a graph of T^2 against n.

Solution 2.6

The graph of T^2 against n, for the data given in solution 2.3, is shown in Figure 2.7. It is approximately a straight line.

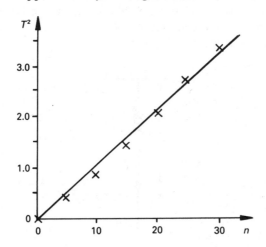

Figure 2.7 A plot of T^2 against n for the oscillations of an elastic string.

It is worthwhile mentioning here that it is quite noticeable that during the oscillation of the spring, the amplitude of the oscillations gradually decreases. This is due to *damping* in the system, which has been ignored. It can be seen from the graphs in Figures 2.2, 2.3 and 2.7 that the effects of damping in the experiment are small, so that the mathematical model of the situation compares well with the experiment. Damping will always be neglected in the systems considered in this book.

Although we have considered the special case of a spring–mass system vibrating under gravity, the equation of motion, equation (2.6), is much more general. It represents any undamped vibrating system with one degree of freedom.

Equation (2.6) and its solution, equation (2.7), involve the displacement from the equilibrium position. Similar equations may be obtained in terms of the displacement from the unstretched position. The relevant equation and its solution are:

$$\ddot{x} + \omega^2 x = g \qquad\qquad (2.10)$$

$$x = \frac{g}{\omega^2} + A \cos(\omega t + \phi) \qquad\qquad (2.11)$$

PFTA 2.7

This exercise concerns the vibration of the spring–mass system described earlier but with the displacement measured from the unstretched position of the spring as shown in Figure 2.8. Show that, using Hooke's law and Newton's second law, the equation of motion may be written in the form $\ddot{x} = -\omega^2 x + g + \omega^2 l$, where $\omega^2 = k/m$.

Figure 2.8 A spring–mass system in motion under gravity.

Solution 2.7

Using the notation in Figure 2.8, and following the procedure which led to equation (2.6), we have:

Hooke's law: $\qquad\qquad\qquad T_x = k(x - l)$

Newton's second law: $\qquad mg - T_x = m\ddot{x}$
$$m\ddot{x} = mg - k(x - l)$$
$$= -kx + mg + kl$$
$$\ddot{x} = -\omega^2 x + g + \omega^2 l \qquad \text{where } \omega^2 = \frac{k}{m}$$

PFTA 2.8

Consider the equation of motion developed in PFTA 2.7. Make the substitution $x = X + l + g/\omega^2$ and show that X satisfies the equation of SHM.

18

Hence deduce the general solution for x.

Solution 2.8

If $x = X + l + g/\omega^2$, then $\dot{x} = \dot{X}$ and $\ddot{x} = \ddot{X}$. Hence, substituting in the differential equation in the solution to PFTA 2.7, we have:

$$\ddot{X} = -\omega^2\left(X + l + \frac{g}{\omega^2}\right) + g + \omega^2 l$$
$$= -\omega^2 X$$

That is, X satisfies the equation of SHM. Hence, using equations (2.6) and (2.8), it follows that $X = A \cos(\omega t + \phi)$ and, finally, we have:

$$x = l + \frac{g}{\omega^2} + A \cos(\omega t + \phi)$$

PFTA 2.9

Interpret the solution to PFTA 2.8.

Solution 2.9

If $X \equiv 0$, then $x = l + g/\omega^2$. Consequently, this is the equilibrium position. We see, then, that the mass, m, oscillates with SHM about the equilibrium position.

PFTA 2.10

In this exercise, you will perform another simple experiment leading to SHM. The *simple pendulum* consists of a length of string which is fixed at one end and has a mass attached at the other. The pendulum is allowed to oscillate freely in a vertical plane. It is important that the angular displacement is small, i.e. the initial displacement must be small. For different

19

lengths, l, of the string, measure the period, T. Plot a graph of T^2 against l, showing that it is approximately a straight line.

Solution 2.10

The following table contains values from a typical experiment:

l(m)	0.92	1.05	1.40	1.56	1.89
T(s)	1.92	2.08	2.38	2.51	2.76

The graph of T^2 against l is shown in Figure 2.9, from which we see that it is approximately a straight line.

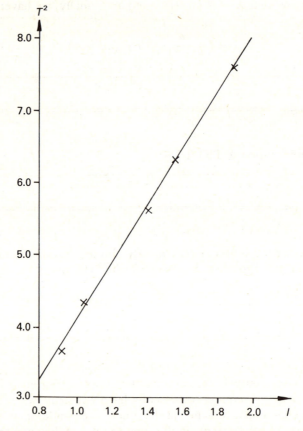

Figure 2.9 T^2 against l for the simple pendulum.

PFTA 2.11

Show that the equation of motion of the simple pendulum is:

$$\ddot{\theta} = -\frac{g}{l} \sin \theta$$

where θ is the angle between the string and the downward vertical. Hence, deduce that, for *small* oscillations, the equation of motion of the pendulum bob is approximately SHM, given by $\ddot{\theta} = -(g/l)\theta$.

Measure the slope of the graph that you obtained in solution 2.10 and hence obtain an approximate value for g, the acceleration due to gravity.

Solution 2.11

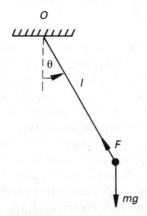

Figure 2.10 A simple pendulum.

Consider the simple pendulum system in Figure 2.10. By taking moments about O:

$$mg \sin \theta = -ml\ddot{\theta}$$

$$\ddot{\theta} = -\frac{g}{l} \sin \theta$$

If θ is small, then $\sin \theta \simeq \theta$ and it follows that the equation of motion of the pendulum bob is approximately:

$$\ddot{\theta} = -\frac{g}{l}\theta$$

21

That is, the motion is approximately SHM with angular frequency $\sqrt{g/l}$ and period given by:

$$T = 2\pi \sqrt{\frac{l}{g}}$$

By squaring both sides, we find that:

$$T^2 = \frac{4\pi^2 l}{g}$$

and hence the relationship between T^2 and l for the simple pendulum is linear. Hence, the slope of the graph of T^2 against l has the value $4\pi^2/g$.

The slope of the graph in the solution to PFTA 2.10 is 4.01. Hence, we conclude that $g = 4\pi^2/4.01 = 9.9 \text{ ms}^{-2}$.

2.2 SYSTEMS WITH TWO DEGREES OF FREEDOM – NORMAL MODES

A system with two degrees of freedom is one which involves two displacements which can be varied independently of one another. To investigate such a system, we can perform the following simple experiment to illustrate what we might expect to happen. The analysis for this experiment is, in fact, a little complicated, so we shall not follow it through. However, the experiment does show some important features.

Consider two identical springs attached to the ends of a rod with the other two ends of the springs fixed so that the system hangs at rest in a vertical plane as shown in Figure 2.11. We can set the system in motion in two different ways: firstly, by pulling both ends down by an equal amount and then releasing the rod (see Figure 2.12(a)); secondly, by pulling one end down and pushing one end up by equal amounts (see Figure 2.12(b)).

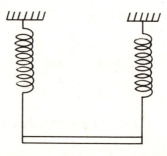

Figure 2.11 Two identical springs with a rod attached to their ends.

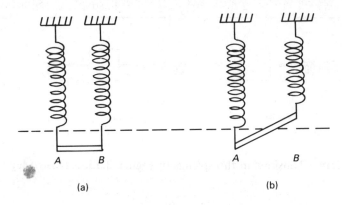

Figure 2.12 Initial positions of the rod.

In the first case, notice that the rod performs SHM in which its position remains horizontal, i.e. the motions of the ends A and B are the same. We say that the motions of A and B are *in phase*. In the second case, however, notice that the rod appears to rotate about its midpoint. Again, the springs perform SHM but in this case the motion of A is exactly opposite to that of B and the angular frequency is different, e.g. when A moves up B moves down. We say that the motions of A and B are *phase opposed*.

These two special motions, starting from the particular initial positions, are called the *normal modes* of vibration of the system. The angular frequencies are known as the *normal mode angular frequencies*; they are the only frequencies at which the system can perform SHM. With any other initial configuration, the resulting motion is certainly not SHM. This can be demonstrated by choosing an arbitrary initial position for the rod and then releasing it.

Now that we know what to expect, we shall consider a system comprising three identical springs, AB, BC and CD, of stiffness k and natural length l, with particles of mass m attached at B and C, and with A and D fixed. The system is free to move along the straight line joining A to D on a smooth horizontal table. The set-up is shown in Figure 2.13, where x_1 and x_2 are the displacements from the equilibrium position.

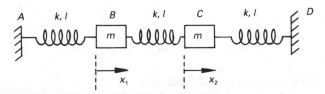

Figure 2.13 A spring–mass system comprising three springs and two masses, each spring having stiffness k and natural length l.

23

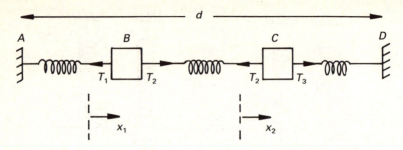

Figure 2.14 Tensions in the springs of Figure 2.13. AD is of length d.

Firstly, suppose that, in the equilibrium position, the tensions in the springs AB, BC and CD are T_1, T_2 and T_3 respectively, as shown in Figure 2.14. Using Hooke's law:

$$T_1 = k(x_1 - l)$$
$$T_2 = k(x_2 - x_1 - l)$$
$$T_3 = k(d - x_2 - l) \qquad \qquad \textbf{(2.12)}$$

In equilibrium, $T_1 = T_2 = T_3$, hence:

$$k(x_1 - l) = k(x_2 - x_1 - l) \quad \text{so that } 2x_1 = x_2$$
$$k(x_2 - x_1 - l) = k(d - x_2 - l) \quad \text{so that } x_1 = 2x_2 - d$$

It follows, then, that in equilibrium:

$$x_1 = \frac{d}{3} \quad \text{and} \quad x_2 = \frac{2d}{3} \qquad \qquad \textbf{(2.13)}$$

Newton's second law for particle B gives:

$$T_2 - T_1 = m\ddot{x}_1$$

Using equations (2.12):

$$kx_2 - kl - 2kx_1 + kl = m\ddot{x}_1$$

which is usually written in the form:

$$m\ddot{x}_1 = -2kx_1 + kx_2 \qquad \qquad \textbf{(2.14)}$$

Newton's second law for particle C gives:

$$T_3 - T_2 = m\ddot{x}_2$$

and hence using equations (2.12):

$$m\ddot{x}_2 = kx_1 - 2kx_2 + kd \tag{2.15}$$

If we now measure displacements from the equilibrium position by writing:

$$X_1 = x_1 - \frac{d}{3} \quad \text{and} \quad X_2 = x_2 - \frac{2d}{3}$$

equations (2.14) and (2.15) then become:

$$m\ddot{X}_1 = -2kX_1 + kX_2$$
$$m\ddot{X}_2 = kX_1 - 2kX_2 \tag{2.16}$$

The equations of motion, equations (2.14) and (2.15), constitute a pair of second-order ordinary differential equations and the solution will yield the motion of the system. However, at the moment, we are interested only in the normal modes. If particles B and C are vibrating in a normal mode, then they each perform SHM with angular frequency ω, say. It is the possible values of ω that we seek.

From equation (2.6):

$$\ddot{X}_1 = -\omega^2 X_1 \quad \text{and} \quad \ddot{X}_2 = -\omega^2 X_2 \tag{2.17}$$

and hence from equation (2.16):

$$-m\omega^2 X_1 = -2kX_1 + kX_2$$
$$-m\omega^2 X_2 = kX_1 - 2kX_2$$

Consequently, we have two values of the ratio X_2/X_1, given by:

$$\frac{X_2}{X_1} = 2 - \frac{m\omega^2}{k} \quad \text{and} \quad \frac{X_2}{X_1} = \frac{1}{2 - m\omega^2/k} \tag{2.18}$$

Thus, eliminating the ratio X_2/X_1:

$$\left(2 - \frac{m\omega^2}{k}\right)^2 = 1$$

Thus $(2 - m\omega^2/k)^2 = \pm 1$ and hence $m\omega^2/k = 2 \pm 1$. Thus:

$$\omega^2 = \frac{k}{m} \quad \text{or} \quad \omega^2 = \frac{3k}{m}$$

That is, the normal mode angular frequencies are:

$$\omega_1 = \sqrt{\frac{k}{m}} \quad \text{and} \quad \omega_2 = \sqrt{\frac{3k}{m}} \tag{2.19}$$

We can find the so-called normal mode displacement ratios from equations (2.18):

(i) When $\omega = \omega_1 = \sqrt{k/m}$, $X_2/X_1 = 1$,
   i.e. $X_2 = X_1$, and the motion is in phase.
(ii) When $\omega = \omega_2 = \sqrt{3k/m}$, $X_2/X_1 = -1$,
   i.e. $X_2 = -X_1$, and the motion is phase opposed.

The two normal modes are illustrated in Figure 2.15.

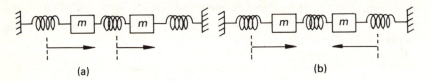

(a)   (b)

Figure 2.15 Normal modes: (a) $\omega_1 = \sqrt{k/m}$, the motion is in phase; (b) $\omega_2 = \sqrt{3k/m}$, the motion is phase opposed.

Normal modes are, of course, very special motions of the system and as such are particularly interesting. However, their usefulness goes much further than that. Suppose the system was set in motion from some arbitrary initial configuration. Certainly, the particles will not vibrate in SHM. However, equation (2.17) gives the equations of motion. (It is convenient to use x_1 and x_2 to measure displacement from the equilibrium position.) In this case, there is a particularly simple method of solution: if the equations are added:

$$m(\ddot{x}_1 + \ddot{x}_2) = -k(x_1 + x_2)$$

and if they are subtracted:

$$m(\ddot{x}_1 - \ddot{x}_2) = -3k(x_1 - x_2)$$

26

If we now write $X_1 = x_1 + x_2$ and $X_2 = x_1 - x_2$, then:

$$m\ddot{X}_1 = -kX_1 \quad \text{and} \quad m\ddot{X}_2 = -3kX_2$$

or:

$$\ddot{X}_1 = -\omega_1^2 X_1 \quad \text{and} \quad \ddot{X}_2 = -\omega_2^2 X_2 \qquad (2.20)$$

Thus:

$$X_1 = A_1 \cos(\omega_1 t + \phi_1) \quad \text{and} \quad X_2 = A_2 \cos(\omega_2 t + \phi_2)$$

Finally, since $x_1 = (X_1 + X_2)/2$ and $x_2 = (X_1 - X_2)/2$, it follows that:

$$x_1 = \frac{A_1 \cos(\omega_1 t + \phi_1) + A_2 \cos(\omega_2 t + \phi_2)}{2}$$

$$x_2 = \frac{A_1 \cos(\omega_1 t + \phi_1) - A_2 \cos(\omega_2 t + \phi_2)}{2} \qquad (2.21)$$

Equations (2.21) describe the motion of the system, from which we can see that the motion of each particle is the sum of two simple harmonic terms, one with the first normal mode angular frequency and the other with the second. For this reason, we often say that the general motion is a *linear combination* of the normal mode motions.

Another system with two degrees of freedom whose normal modes are easy to visualise, but a little harder to develop mathematically, is that of two particles of equal mass m performing transverse vibrations on a light elastic string. Suppose that the string is of length l and that the masses are equidistant from the fixed ends of the string. We assume that the tension, T, in the string is sufficiently large so as not to vary during the motion.

PFTA 2.12

Consider the transverse vibrations of the two masses on a light elastic string shown in Figure 2.16. The tension T is considered to be constant and the displacements x_1 and x_2 are small. Show that $\cos\theta_1 \simeq \cos\theta_2 \simeq \cos\theta_3 \simeq 1$, and that $\sin\theta_1 \simeq 3x_1/l$, $\sin\theta_2 \simeq 3(x_2 - x_1)/l$ and $\sin\theta_3 \simeq 3x_2/l$.

Use Newton's second law to obtain the equations of motion in the form:

$$T \sin\theta_2 - T \sin\theta_1 = m\ddot{x}_1$$

$$-T \sin\theta_3 - T \sin\theta_2 = m\ddot{x}_2$$

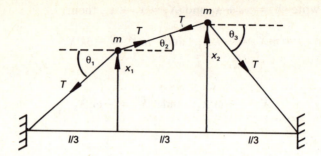

Figure 2.16 Two masses attached to light elastic strings performing transverse vibrations.

Hence show that:

$$m\ddot{x}_1 = \frac{3T(-2x_1 + x_2)}{l}$$

$$m\ddot{x}_2 = \frac{3T(x_1 - 2x_2)}{l}$$

Finally, by setting $\ddot{x}_1 = -\omega^2 x$ and $\ddot{x}_2 = -\omega^2 x$, show that the equations of motion may be written in the form:

$$\left(2 - \frac{ml\omega^2}{3T}\right)x_1 - x_2 = 0$$

$$-x_1 + \left(2 - \frac{ml\omega^2}{3T}\right)x_2 = 0$$

In a normal mode of vibration, we require non-zero values for x_1 and x_2. Show that this leads to the equation:

$$\left(2 - \frac{ml\omega^2}{3T}\right)^2 - 1 = 0$$

and hence show that the two normal mode angular frequencies are given by:

$$\omega_1 = \sqrt{\frac{3T}{ml}} \quad \text{and} \quad \omega_2 = \sqrt{\frac{9T}{ml}}$$

The displacements in a normal mode may be written in the form $x_2 = Rx_1$, where R is called the *normal mode displacement ratio*. Show that in the first mode $R = 1$ and in the second mode $R = -1$.

Solution 2.12

Since θ_1, θ_2 and θ_3 are small, we can say that $\cos \theta_1 \simeq \cos \theta_2 \simeq \cos \theta_3 \simeq 1$. Also, $\tan \theta_1 = x_1/(l/3)$ and since θ_1 is small, it follows that $\sin \theta_1 \simeq \tan \theta_1 = 3x_1/l$. Similarly, we find that $\sin \theta_2 \simeq 3(x_2 - x_1)/l$ and $\sin \theta_3 \simeq 3x_2/l$.

The equations of motion of the two masses are, using Newton's second law:

$$T \sin \theta_2 - T \sin \theta_1 = m\ddot{x}_1$$

$$-T \sin \theta_3 - T \sin \theta_2 = m\ddot{x}_2$$

Using the approximate values, we find that:

$$\frac{3T(x_2 - x_1)}{l} - \frac{3Tx_1}{l} = m\ddot{x}_1$$

$$\frac{3Tx_2}{l} - \frac{3T(x_2 - x_1)}{l} = m\ddot{x}_2$$

Hence, after rearranging, it follows that:

$$m\ddot{x}_1 = \frac{3T(-2x_1 + x_2)}{l}$$

$$m\ddot{x}_2 = \frac{3T(x_1 - 2x_2)}{l}$$

In a normal mode of vibration, $\ddot{x}_1 = -\omega^2 x_1$ and $\ddot{x}_2 = -\omega^2 x_2$ Hence:

$$-\omega^2 x_1 = \frac{3T(-2x_1 + x_2)}{ml}$$

$$-\omega^2 x_2 = \frac{3T(x_1 - 2x_2)}{ml} \tag{2.22}$$

If we multiply by $ml/3T$ and rearrange, we find that:

$$\left(2 - \frac{ml\omega^2}{3T} \right) x_1 - x_2 = 0$$

$$-x_1 + \left(2 - \frac{ml\omega^2}{3T} \right) x_2 = 0$$

For non-zero values of x_1 and x_2, we require that:

$$\left(2 - \frac{ml\omega^2}{3T}\right)^2 - 1 = 0$$

Hence:

$$2 - \frac{ml\omega^2}{3T} = \pm 1$$

so that:

$$\omega^2 = \frac{(2 \pm 1)3T}{ml}$$

$$= \frac{3T}{ml} \quad \text{or} \quad \frac{9T}{ml}$$

That is, the normal mode angular frequencies are:

$$\omega_1 = \sqrt{\frac{3T}{ml}} \quad \text{and} \quad \omega_2 = \sqrt{\frac{9T}{ml}}$$

With $\omega_1 = 3T/ml$, equations (2.22) lead to:

$$x_1 - x_2 = 0 \quad \text{and} \quad -x_1 + x_2 = 0$$

which gives $x_2 = -x_1$. Hence, the normal mode displacement ratio is given by $R = 1$.

With $\omega_2 = 9T/ml$, equations (2.22) lead to:

$$-x_1 - x_2 = 0 \quad \text{and} \quad -x_1 - x_2 = 0$$

which gives $x_2 = -x_1$. Hence, the normal mode displacement ratio is given by $R = -1$. The normal mode configurations are shown in Figure 2.17.

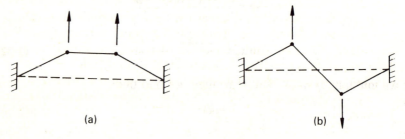

(a)   (b)

Figure 2.17 Normal mode configurations for two equal masses on an elastic string: (a) $\omega_1 = \sqrt{3T/ml}$, the motion is in phase; (b) $\omega_2 = \sqrt{9T/ml}$, the motion is phase opposed.

2.3 SYSTEMS WITH MORE THAN TWO DEGREES OF FREEDOM

In principle, it is not difficult to extend the ideas developed in the last section to cope with systems with three or more degrees of freedom. The mathematics, however, becomes much more involved. For three identical masses equally distributed on a string, we could use a similar, but much more tedious, approach to show that:

$$\omega_1^2 = \frac{(2 - \sqrt{2})4T}{ml} \qquad \omega_2^2 = \frac{8T}{ml} \qquad \omega_3^2 = \frac{(2 + \sqrt{2})4T}{ml}$$

The normal modes are shown in Figure 2.18.

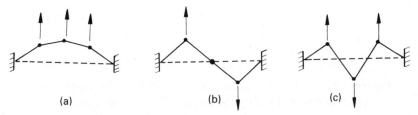

(a)   (b)   (c)

Figure 2.18 Normal modes for three masses on a string. Notice that in (b) the centre mass remains at rest.

In these examples, we have assumed that the string is light, i.e. it has no mass. Now consider the problem of transverse vibrations of a string of length l whose mass per unit length is ρ and the total mass of the string is ρl. It is usual to introduce the quantity c given by $c^2 = T/\rho$.

One way in which we may approximate this string is as a sequence of identical masses equally distributed along the string. The sum of the identical masses must be the total mass of the string, and the more masses we introduce the better the approximation. The string is said to be a *continuous parameter* system, whereas the approximation to it is called a *lumped parameter* system, e.g. if we use five equal masses equally distributed along the string with one mass at each end, the normal modes of the lumped parameter system are as shown in Figure 2.18.

We would get a better approximation to the continuous string if we use, say, 12 masses, each with $m = \rho l/12$, and place them at distances $l/11$ apart as shown in Figure 2.19.

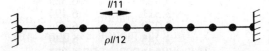

$l/11$

$\rho l/12$

Figure 2.19 A lumped parameter approximation, with 12 masses, to a uniform continuous string.

31

With this set-up, it can be shown that the angular frequencies of the first three normal modes are given by:

$$\omega_1 = \frac{3.270c}{l} \qquad \omega_2 = \frac{6.474c}{l} \qquad \omega_3 = \frac{9.546c}{l}$$

The corresponding mode shapes are shown in Figure 2.20.

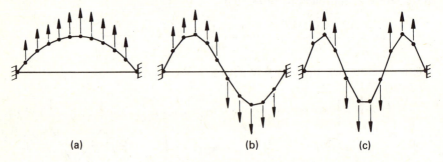

(a)   (b)   (c)

Figure 2.20   First three normal modes for the 12-lumped mass approximation for the transverse vibrations of a string.

Thus, we have approximations to the first three normal mode angular frequencies for the transverse vibrations of a string. Such vibrations will be considered in Chapter 4, where we shall see that the values for the continuous string are $\omega_1 = 3.141c/l$, $\omega_2 = 6.283c/l$ and $\omega_3 = 9.425c/l$. The values of the approximation to the angular frequencies of the first three normal modes are shown in Table 2.1. From the data in Table 2.1, we can see that our approximations with 12 masses are quite good, the errors in the calculation of the angular frequencies being approximately 4%, 3% and 1% respectively.

Table 2.1   A comparison of the angular frequencies of the first three normal modes ranging from the five-lumped parameter approximation to the continuous string.

Number of masses on string	$l\omega_1/c$	$l\omega_2/c$	$l\omega_3/c$
5	3.423	6.325	8.263
9	3.311	6.494	9.428
12	3.270	6.474	9.546
15	3.245	6.449	9.572
continuous string	3.141	6.283	9.425

3 WAVE MOTION

3.1 PROGRESSIVE WAVES

Consider a very long string stretched along the x-axis, the string being at
rest. Suppose that one end is held and given a sharp 'flick', setting it in
motion. The disturbance set up by flicking the string will travel down the
string. During the passage of the disturbance, each particle of the string
remains at rest until the disturbance reaches it, at which instant the particle
is displaced, returning to its original position once the disturbance has
passed.

If we assume that the disturbance propagates *without change of shape*,
then a photograph taken of the disturbance at two different times would
show exactly the same picture. Such a picture is called the *wave profile*.

Suppose that the wave propagates with velocity c and that initially the
equation of the disturbance is $y = f(x)$, as shown in Figure 3.1, where f is
an arbitrary function of one variable. Notice that the wave propagates

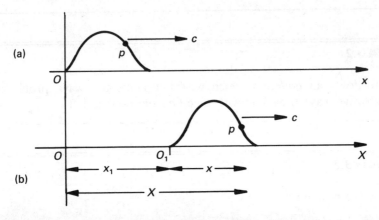

Figure 3.1   (a) Transverse disturbance on a string at time $t = 0$. (b)
Wave profile at time t.

33

along the direction of the string, whereas the disturbance of each particle is perpendicular to the string. Such a wave is called a *transverse wave*.

At some time, t, later the wave profile would have moved a distance $x_1 = ct$ as shown in Figure 3.1(b). Referring to the origin O_1, for X, the wave profile at time t must be $y = f(X)$, because the wave propagates without change of shape. However, it is clear that $x = x_1 + X$. Hence, $X = x - ct$. Thus, at time t the wave profile has equation:

$$y = f(x - ct) \qquad (3.1)$$

This equation is extremely important in the mathematical description of wave motion. Such a wave is called a *progressive wave* (or *travelling wave*) and it moves with speed c from left to right. Note that it is the combination $x - ct$ and functions of this combination that are possible progressive wave profiles.

PFTA 3.1

Give two examples of possible progressive wave profiles and two examples which are not possible progressive wave profiles.

Solution 3.1

$(x - ct)^2$ and $\cos(x - ct)$ are possible progressive wave profiles, whereas $x^2 - c^2t^2$ and $\cos x - \cos ct$ are not.

PFTA 3.2

Write down an equation which would describe the wave profile of a progressive wave travelling with speed c from right to left.

Solution 3.2

The wave profile is:

$$y = g(x + ct) \qquad (3.2)$$

where g is an arbitrary function of one variable. This follows immediately from equation (3.1) since motion with velocity c from right to left is equivalent to motion with velocity $-c$ from left to right.

The mathematical equation which describes the motion of a wave is called the *wave equation*, and equations (3.1) and (3.2) are said to be solutions of the wave equation.

Equations (3.1) and (3.2) are examples of functions of two variables, so for a detailed analysis of wave motion, we need to be able to handle the mathematics of such functions. But such mathematics is beyond the scope of this text. However, a good indication of the mathematical description is possible using only the calculus and algebra covered in a sixth form course. Consequently, we shall not write down the wave equation explicitly.

Return now to equation (3.1) and consider a particular point on the wave profile – say, for example, the point P in Figure 3.1(a). The point P has a fixed y coordinate so that at P, $f(x - ct) =$ constant for all x and t. Hence, the point P must be such that $x - ct$ is constant, the value of the constant depending on which part of the profile that P is situated. Thus, if $x - ct =$ constant, differentiating with respect to time will give:

$$\frac{dx}{dt} - c = 0 \quad \text{or} \quad \frac{dx}{dt} = c$$

The point P is often referred to as a particular phase of the wave, and since dx/dt is the velocity of P, it is called the *phase velocity*. Consequently, the progressive wave has phase velocity c. Clearly, for a wave travelling from right to left the phase velocity is $-c$.

3.2 HARMONIC WAVES

So far, we have described the wave profile in terms of arbitrary functions, f and g, as in equations (3.1) and (3.2). One of the simplest cases is when the profile at time $t = 0$ is a simple cosine curve $y = \cos mx$. This leads to the wave profile at time t as:

$$y = A \cos m(x - ct) \tag{3.3}$$

Such a wave is called a *harmonic wave*, or sometimes a *sinusoidal wave*, and it is a wave moving to the right with velocity c. Similarly, $y = A \cos m(x + ct)$ is a harmonic wave moving to the left with velocity c.

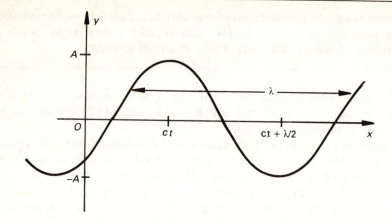

Figure 3.2 The wave profile $y = A \cos m(x - ct)$ with $m = 2\pi/\lambda$.

Since the harmonic wave profile in equation (3.3) is periodic, we would expect such waves to be oscillatory in nature, and indeed that is the case. If we compare equation (3.3) with equation (2.7) for the vibrating spring, we see that the form of the displacement is very similar. The consequence of this is that the terminology for the parameters of a wave profile follows from that for SHM.

The quantity A is called the *amplitude* of the wave.

The wave profile repeats itself at distances $2\pi/m$ (see Figure 3.2). The quantity $\lambda = 2\pi/m$ is called the *wavelength* of the profile and equation (3.3) may be written as:

$$y = A \cos \frac{2\pi}{\lambda} (x - ct) \qquad (3.4)$$

The time taken to pass through one complete cycle is called the *period*, T. Since the cosine function is periodic with period 2π, it follows that $2\pi cT/\lambda = 2\pi$, i.e. $T = \lambda/c$.

The *frequency*, f, of the harmonic wave is the number of waves passing a fixed point in unit time. This is related to the period by:

$$f = \frac{1}{T}$$

so that the relationship between velocity, frequency and wavelength is:

$$c = f\lambda \qquad (3.5)$$

36

The quantity $k = 1/\lambda$, the number of waves in unit distance, is called the *wave number*.

Finally, the *angular frequency* is given by $\omega = 2\pi/T = 2\pi f$.

We thus have four equivalent forms of the harmonic wave profile in equation (3.3):

$$y = A \cos 2\pi \left(\frac{x}{\lambda} - \frac{t}{T} \right)$$

$$y = A \cos 2\pi \left(\frac{x}{\lambda} - ft \right)$$

$$y = A \cos 2\pi (kx - ft)$$

$$y = A \cos (2\pi kx - \omega t) \tag{3.6}$$

Now consider a second wave profile:

$$y = A \cos [2\pi(kx - ft) + \phi]$$

We see that this is just the wave profile in equation (3.6) but displaced a distance $\phi/2\pi k$. ϕ is called the *phase* of the profile and, just as was the case in Chapter 2, if $\phi = 0, 2\pi, 4\pi, \ldots$ then the two profiles are *in phase*, while if $\phi = \pi, 3\pi, 5\pi, \ldots$ then the two profiles are *phase opposed*.

PFTA 3.3

Show that the wave profile $y = A \cos [m(x - ct) + \phi]$ may be written as the profile $y = A \sin [m(x - ct) + \psi]$. What is the relationship between the two phases ϕ and ψ?

Solution 3.3

Since $\cos \alpha = \sin (\alpha + \pi/2)$, it follows that:

if $y = A \cos [m(x - ct) + \phi]$ then $y = A \sin [m(x - ct) + \phi + \pi/2]$

That is, $y = A \sin [m(x - ct) + \psi]$ is an equivalent harmonic wave profile with $\psi = \phi + \pi/2$.

It is possible for the phase of a wave to change at a point of reflection.

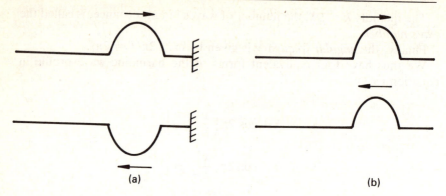

Figure 3.3 Reflection of a pulse on a string: (a) fixed end with a change of phase; (b) free end with no change of phase.

We can demonstrate this by a simple experiment.

Take a long piece of string and 'flick' one end. This will result in a pulse being sent down the string. If the other end is free, then the pulse will be reflected the same way up. However, if the other end is fixed, the pulse will be reflected upside-down, which means that the reflected wave differs in phase by 180° from the incident wave. The two cases are illustrated in Figure 3.3.

The reason for the change of phase is as follows. When the pulse arrives at the fixed end, it exerts a force on the rigid support. By Newton's third law, the support exerts an equal and opposite force on the string, and this reaction generates a pulse which travels back along the string with its profile reversed.

3.3 SUPERPOSITION

It is an experimental observation that, for many different kinds of wave, two or more waves may propagate independently in the same medium. This means that the displacement of each particle in the medium at a given time is the sum of the displacements due to each of the wave profiles propagating through the medium. This process is called the *principle of superposition*; it is an extremely important process in applied mathematics. Systems which obey the superposition principle are called *linear systems*, while the mathematical equations that describe them are said to be linear equations. The importance of the principle is that a complicated wave motion can be analysed in terms of its simple, individual constituent waves. In fact, it was shown by the French mathematician Joseph Fourier that any periodic motion can be expressed as the superposition of harmonic waves. The most general expression of such a combination is called a *Fourier series*, which we shall discuss in Chapter 4.

An interesting combination is that of waves travelling with the same speed but in opposite directions. The most general combination is known as *D'Alembert's* solution of the wave equation:

$$y = f(x - ct) + g(x + ct) \qquad (3.7)$$

where f and g are arbitrary functions.

Now let us consider the superposition of two harmonic waves with the same frequency:

$$y_1 = A_1 \cos (2\pi kx - ct)$$
$$y_2 = A_2 \cos (2\pi kx - ct + \phi)$$

The superposition of these two waves can be shown to be:

$$y_1 + y_2 = A \cos (2\pi kx - ct + \psi)$$

where:

$$A = (A_1^2 + A_2^2 + 2A_1A_2 \cos \phi)^{1/2} \quad \text{and} \quad \tan \psi = \frac{A_2 \sin \phi}{A_1 + A_2 \cos \phi} \qquad (3.8)$$

This wave travels with the same velocity and has the same frequency as the two constituent waves but has an amplitude A given by equation (3.8). If the phase difference between waves y_1 and y_2 is small, then $\cos \phi \simeq 1$ and $\sin \phi \simeq 0$, hence:

$$A \simeq (A_1^2 + A_2^2 + 2A_1A_2)^{1/2} = A_1 + A_2 \quad \text{and} \quad \tan \psi \simeq 0 \qquad \text{(i.e. } \psi = 0\text{)}$$

This means that the resulting wave has an amplitude which is approximately the sum of the two amplitudes and is in phase with y_1 (see Figure 3.4).

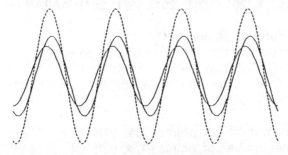

Figure 3.4 Superposition of two wave profiles that are nearly in phase, $t = 0$.

PFTA 3.4

y_1 is the wave profile given by $y_1 = A \cos (2\pi kx - ct)$. Find the superposition, $y_1 + y_2$, for the following wave profiles, y_2: (i) $y_2 = y_1$; (ii) $y_2 = A \cos (2\pi kx + ct)$; (iii) $y_2 = A \cos (2\pi kx - ct + \phi)$; (iv) $y_2 = B \cos (2\pi kx - ct + \phi)$.

Solution 3.4

(i) $y_1 + y_2 = 2A \cos (2\pi kx - ct)$.
(ii) $y_1 + y_2 = A[\cos (2\pi kx - ct) + \cos (2\pi kx + ct)] = 2A \cos 2\pi kx \cos ct$.
(iii) $y_1 + y_2 = 2A \cos 2\pi kx \cos (ct - \phi/2)$.
(iv) $y_1 + y_2 = A \cos (2\pi kx - ct) + B \cos (2\pi kx - ct + \phi)$
$= (A + B \cos \phi) \cos (2\pi kx - ct) - B \sin \phi \sin (2\pi kx - ct)$.

The following segment of pseudo-code provides a procedure to illustrate the superposition of the two profiles:

$$y_1 = A_1 \cos (2\pi k_1 x + \phi_1) \quad \text{and} \quad y_2 = A_2 \cos (2\pi k_2 x + \phi_2)$$

The superposed wave is $y_3 = y_1 + y_2$.

```
PROCEDURE superposition (A1, k1, ph1, A2, k2, phi2: REAL)
CONSTANT
    two_pi := 6.283185308;
    number_of_points := 200;
VAR
    i: INTEGER;
    x[1. .201], y1[1. .201], y2[1. .201], y3[1. .201]: ARRAY OF REAL;
BEGIN
    DO i = 1, number_of_points+1
        x[i] := FLOAT(i-1)/FLOAT(number_of_points/2);
        y1[i] := A1*COS(two_pi*k1*x[i]+phi1);
        y2[i] := A2*COS(two_pi*k2*x[i]+phi2);
        y3[i] := y1[i]+y2[i];
    END DO;
    set_polyline(number_of_points+1, x, y1);
    set_polyline(number_of_points+1, x, y2);
    set_polyline(number_of_points+1, x, y3);
END;
```

Choose the graphics window to be defined by the rectangle $(0, -1)$, $(2, -1)$, $(2, 1)$ and $(0, 1)$. If values of A1 and A2 are chosen such that $0 < A1 + A2 < 1$, then the resulting superposed wave will lie inside the window.

3.4 STANDING WAVES

Consider the superposition:

$$y = A \cos (2\pi kx - ct) + A \cos (2\pi kx + ct)$$

which is shown in PFTA 3.4 to be given by:

$$y = 2A \cos 2\pi kx \cos ct$$

If we write $A_0 = 2A$, the profile is given by:

$$y = A_0 \cos 2\pi kx \cos ct \qquad (3.9)$$

This is known as a *stationary wave*, or *standing wave*, the name coming from the fact that the wave profile does not move forward.

PFTA 3.5

Show that the standing wave described by equation (3.9) does not move forward.

Solution 3.5

The standing wave has profile $y = A_0 \cos 2\pi kx \cos ct$. Consider any fixed point on the string, say $x = X$. Then the displacement at X is given by $y = A_0 \cos 2\pi kX \cos ct$ which may be written as $y = B \cos ct$ where $B = A_0 \cos 2\pi kX$. Since B is a constant, the motion of that particle of the string at $x = X$ is given by $y = B \cos ct$. Hence, it follows that this particle performs SHM, i.e. the particle simply moves from side to side. Since this is true for all fixed points X, it follows that the wave is a side-to-side motion and consequently it does not propagate forwards.

At points given by $x = 1/4k, 3/4k, 5/4k, \ldots$ the standing wave displacement given by equation (3.9) always vanishes, i.e. $y = 0$. Such points are

41

called *nodes*. Mid-way between the nodes are points called *anti-nodes*. At these points, the magnitude of the displacement is a maximum and is given by $2A|\cos ct|$.

PFTA 3.6

Show that the magnitude of the displacement at an anti-node is given by $2A|\cos ct|$.

Solution 3.6

From equation (3.9), $|y| = 2A|\cos 2\pi kx||\cos ct|$. Now at an anti-node $x = 0$, $k/2$, k, $3k/2$, . . . so that $|\cos 2\pi kx| = 1$. Hence, $|y| = 2A|\cos ct|$ at an anti-node.

Standing waves can be generated by the superposition of incident and reflected waves, since this produces the sum of two wave profiles moving in opposite directions.

3.5 WAVES IN TWO AND THREE DIMENSIONS

Good examples of waves in two dimensions are waves seen on the surface of the water in a swimming pool; the waves on the surface are two-dimensional. Many pools now have a 'wave making' machine which causes a wave to travel down the length of the pool. This wave is in the form of a straight line across the width of the pool. This line is known as the *wavefront*, i.e. it is the wavefront which propagates through the medium. In general, for any straight-line wave the situation is as shown in Figure 3.5.

PFTA 3.7

Using the notation of Figure 3.5, show that the equation of the wavefront may be written in the form $lx + my = d$, where $l = \cos \alpha$ and $m = \sin \alpha$.

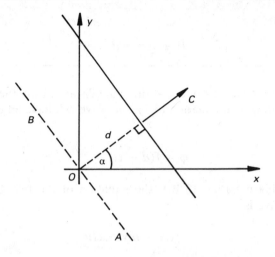

Figure 3.5 A straight-line wave propagating in the *xy* plane. *AB* is a fixed line parallel to the wavefront with slope −*l*/*m*.

Solution 3.7

The perpendicular from O intersects the wavefront at the point Q ($d \cos \alpha$, $d \sin \alpha$). The slope of this perpendicular is m/l, hence the slope of the wavefront is $-l/m$. Hence, the equation of the wavefront may be written as:

$$y = -\frac{l}{m}x + c$$

Since the wavefront passes through Q, $d \sin \alpha = -(l/m) d \cos \alpha + c$. Hence:

$$c = d\left(\sin \alpha + \frac{l}{m}\cos \alpha\right)$$

$$= \frac{d}{m}\,(\sin^2 \alpha + \cos^2 \alpha)$$

$$= \frac{d}{m}$$

It follows that:

$$y = -\frac{l}{m}x + \frac{d}{m}$$

that is:

$$lx + my = d \qquad \textbf{(3.10)}$$

All points on the wavefront are at the same distance from the fixed line. Hence, by analogy with equation (3.1), the wave will have an equation of the form:

$$\phi = f(d - ct)$$

and hence, using equation (3.10), the equation of the two-dimensional straight-line wave is:

$$\phi = f(lx + my - ct) \qquad \textbf{(3.11)}$$

Similarly, a straight-line wave travelling in the opposite direction has the form:

$$\phi = g(lx + my - ct)$$

Of course, we can also use the superposition principle to combine any number of such waves.

Another simple two-dimensional wave is the circular wave, e.g. a ripple on a pond caused by dropping a pebble.

PFTA 3.8

What is the profile of a circular wave?

Solution 3.8

Suppose that r is the radial distance of the circular wavefront from the origin, as shown in Figure 3.6. Since the circular wave propagates radially outwards from the origin, the expression $r - ct$ will play the same role in the circular wave profile as the expression $x - ct$ does in the wave profile for a one-dimensional wave, as given in equation (3.1). Hence, a circular wave has the profile:

$$\phi = f(r - ct) \qquad \textbf{(3.12)}$$

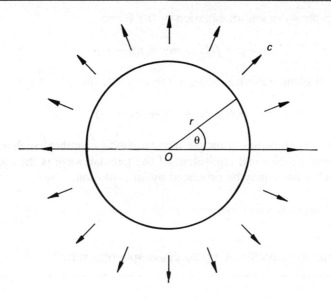

Figure 3.6   A circular wave, where r and θ are polar coordinates.

The three-dimensional analogue of the straight-line wave is the plane wave in which the disturbance is distributed over a plane in three dimensions and propagates along the normal to the surface (see Figure 3.7). The equation of the plane is of the form:

$$lx + my + nz = d$$

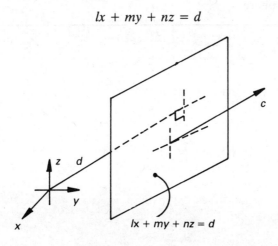

Figure 3.7   A plane wave.

and the plane wave has an equation of the form:

$$\phi = f(lx + my + nz - ct) \tag{3.13}$$

Similarly, a plane wave travelling in the opposite direction has the form:

$$\phi = g(lx + my + nz + ct)$$

Again, the superposition principle may be used to combine such waves.

The three-dimensional equivalent of the circular wave is the spherical wave. Such a wave may be produced by an explosion.

PFTA 3.9

What is the wave profile of a progressive spherical wave?

Solution 3.9

By a direct analogy with solution 3.8, it follows that the wave profile for a spherical wave is given by:

$$\phi = f(r - ct)$$

where r is the distance of the spherical wavefront from the origin at the centre of the sphere.

4 WAVES ON STRINGS AND WAVES IN SPRINGS

4.1 TRANSVERSE WAVES ON STRINGS

In Chapter 3, we discussed progressive and stationary waves by using the example of a wave on a string. In this section, we shall consider waves on strings in more detail. The two major assumptions are that the string is uniform with mass per unit length, ρ, and that the vibrations are sufficiently small so that the tension, T, remains constant.

If we produce a wave on a string, the wave propagates along the direction of the string whereas each particle of the string vibrates from side to side, i.e. perpendicular to the direction of propagation. Consequently, waves on strings are transverse waves.

Many musical instruments, e.g. the piano, the guitar, the violin etc, comprise strings which are fixed at each end and the musical sound is developed by producing vibrations in the strings.

We assume that the amplitude of the vibration is small compared with the length of the string. This is a reasonable assumption in many practical applications, e.g. a violin string is about 30 cm long and the amplitude is less than 2 mm, so the length of the string is more than 150 times as large as the amplitude. In this case, the change in the tension in the string due to the propagation of the wave is small and consequently we can assume that the tension remains constant throughout the motion. The following exercises develop the relationship between the tension, T, and the velocity of propagation, c.

PFTA 4.1

Suppose that we have a small pulse propagating from left to right on a string. Now suppose that the string is pulled from right to left with the same speed so that the pulse itself remains fixed in space. The situation is illustrated in Figure 4.1. Show that the total force acting on the segment,

47

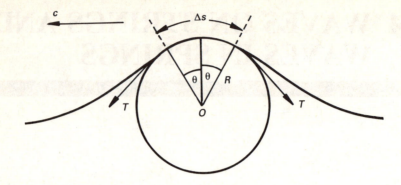

Figure 4.1 The top section of a pulse moving from left to right. The string is moving from right to left with speed c. The small section Δs is approximately the arc of a circle of radius R.

directed towards O, is given by $2T \sin \theta$ and hence show that if θ is small then this is approximately $T\Delta s/R$.

Solution 4.1

Resolving the tensions vertically, we see that the total force on the segment, directed towards O, is:

$$2T \cos \left(\frac{\pi}{2} - \theta \right) = 2T \sin \theta$$

For small vibrations, θ is small and hence $\sin \theta \simeq \theta$. It follows, then, that the total force is approximately:

$$2T\theta = 2T \frac{\Delta s}{2R} \quad \left(\text{since } 2\theta = \frac{\Delta s}{R} \right)$$

$$= T\frac{\Delta s}{R}$$

PFTA 4.2

The small segment of string described in PFTA 4.1 can be considered to be, instantaneously, a particle moving in a circle of radius R. Write down

48

the acceleration of this particle and hence, using Newton's second law and the solution of PFTA 4.1, show that:

$$T\frac{\Delta s}{R} = \rho c^2 \frac{\Delta s}{R}$$

Hence, deduce that:

$$c = \sqrt{\frac{T}{\rho}}$$

Solution 4.2

Since the particle is moving with speed c in a circle of radius R, the acceleration towards O is given by c^2/R. The length of the segment is Δs so that its mass is $\rho\Delta s$. Consequently, Newton's second law and the solution of PFTA 4.1 give:

$$T\frac{\Delta s}{R} = \rho\Delta s\frac{c^2}{R}$$

Hence, it follows that:

$$c = \sqrt{\frac{T}{\rho}}$$

Suppose that we have a string stretched along the x-axis from $x = 0$ to $x = l$ which is set in motion by some mechanism. Such a motion will be a standing wave as described in Chapter 3. At any fixed time, t, the profile will be of the form $y = f(x)$ for a suitable function $f(x)$ (see Figure 4.2).

If we consider harmonic waves, then all points on the string perform SHM with angular frequency ω – the function $f(x)$ includes a sinusoidal term – and hence the wave form has the equation:

$$y = \phi(x) \cos(\omega t + \alpha) \tag{4.1}$$

The amplitude depends on the position, x, and is given by the function $\phi(x)$ which, to ensure that the string has fixed ends, has the property:

$$\phi(0) = \phi(l) = 0$$

49

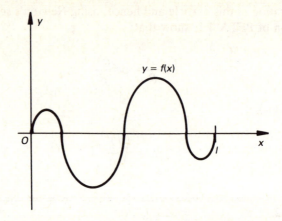

Figure 4.2 Transverse vibration of a stretched string at a fixed time.

These conditions are called *boundary conditions*. In the following exercises, we set up an equation from which we can determine $\phi(x)$.

PFTA 4.3

Consider the motion of a small segment of vibrating string as shown in Figure 4.3. Each point of the string is vibrating in SHM with angular frequency ω. Use Newton's second law and equation (4.1) to show that:

$$T[\sin (\theta + \Delta\theta) - \sin \theta] = -\rho\Delta x\omega^2 y$$

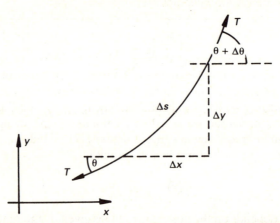

Figure 4.3 A segment of string of length Δs which, when the string is horizontal, is of length Δx so that its mass is $\rho\Delta x$.

Solution 4.3

Newton's second law gives:

$$T \sin (\theta + \Delta\theta) - T \sin \theta = \rho\Delta x \frac{d^2y}{dt^2}$$

Since each point is vibrating in a transverse manner with SHM, the displacement y is given by equation (4.1), hence we may write $d^2y/dt^2 = -\omega^2 y$ and it follows that:

$$T[\sin (\theta + \Delta\theta) - \sin \theta] = -\rho\Delta x \omega^2 y$$

PFTA 4.4

Use the fact that, for small angles, $\sin \theta \simeq \tan \theta = \left(\dfrac{dy}{dx}\right)_x$ to show that

the solution of PFTA 4.3 becomes:

$$\frac{\left(\dfrac{dy}{dx}\right)_{x+\Delta x} - \left(\dfrac{dy}{dx}\right)_x}{\Delta x} = -\frac{\omega^2 y}{c^2}$$

Take the limit as $\Delta x \to 0$ to show that the left-hand side becomes:

$$\frac{d^2y}{dx^2}$$

and hence use equation (4.1) to deduce that:

$$\frac{d^2\phi}{dx^2} = -\frac{\omega^2\phi}{c^2}$$

Solution 4.4

Since θ is small, it follows that:

$$\sin \theta \simeq \tan \theta = \left(\frac{dy}{dx}\right)_x \quad \text{and} \quad \sin (\theta + \Delta\theta) \simeq \tan (\theta + \Delta\theta) = \left(\frac{dy}{dx}\right)_{x+\Delta x}$$

Also, $\rho/T = 1/c^2$, hence the solution of PFTA 4.3 may be written as:

$$\left(\frac{dy}{dx}\right)_{x+\Delta x} - \left(\frac{dy}{dx}\right)_x = -\frac{\omega^2 \Delta x y}{c^2}$$

Dividing by Δx gives the required result.

If we consider the limit as $\Delta x \to 0$, the left-hand side becomes d^2y/dx^2 and hence:

$$\frac{d^2y}{dx^2} = -\frac{\omega^2 y}{c^2}$$

Now using equation (4.1):

$$\frac{d^2y}{dx^2} = \frac{d^2\phi}{dx^2} \cos (\omega t + \alpha)$$

Finally, then, it follows that $\phi(x)$ satisfies the differential equation:

$$\frac{d^2\phi}{dx^2} = -\frac{\omega^2 \phi}{c^2}$$

If we write this as $d^2\phi/dx^2 = -p^2\phi$, where $p = \omega/c$, we can use the solution of PFTA 2.5 to write the solution of this equation in the form:

$$\phi = A \cos (px + \varepsilon) \qquad \text{where } p = \frac{\omega}{c} \qquad (4.2)$$

PFTA 4.5

Show that to satisfy the boundary conditions, we require that:

$$\varepsilon = \frac{\pi}{2} \quad \text{and} \quad \cos\left(pl + \frac{\pi}{2}\right) = 0$$

Hence deduce that $\omega = \omega_n = n\pi c/l$, where $n = 1, 2, 3, \ldots$

Solution 4.5

The boundary conditions require that $\phi(0) = \phi(l) = 0$ so that:

$$A \cos \varepsilon = 0 \quad \text{and} \quad A \cos (pl + \varepsilon) = 0$$

Since $A \neq 0$, the first equation yields $\cos \varepsilon = 0$ and hence we may take $\varepsilon = \pi/2$. The second equation then yields:

$$\cos\left(pl + \frac{\pi}{2}\right) = 0$$

The possible solutions are odd multiples of $\pi/2$, so that:

$$pl + \frac{\pi}{2} = (2n + 1)\frac{\pi}{2} \qquad n = 0, 1, 2, \ldots$$

Now $p = \omega/c$, consequently ω can take any one of the values:

$$\omega_n = \frac{n\pi c}{l} \qquad n = 1, 2, 3, \ldots$$

PFTA 4.6

Why do we reject the solution with $n = 0$ in PFTA 4.5?

Solution 4.6

We reject the case $n = 0$ since this corresponds to $\phi(x) \equiv 0$. This solution satisfies the boundary conditions and the equation but represents the uninteresting case of the string remaining at rest along the x-axis.

The solution $\phi(x) \equiv 0$ is often called the *trivial solution*.

The solution of PFTA 4.5 means that the string can vibrate in a manner such that every particle executes SHM but only with particular frequencies, called the *normal mode frequencies* or the *natural frequencies*. These frequencies are given by:

$$f_n = \frac{\omega_n}{2\pi} = \frac{nc}{2l}$$

Since $T = \rho c^2$, it follows that:

$$f_n = \frac{n}{2l}\sqrt{\frac{T}{\rho}} \tag{4.3}$$

Equation (4.3) is known as *Mersenne's law*.

The lowest natural frequency is called the *fundamental* and is given by:

$$f_1 = \frac{1}{2l}\sqrt{\frac{T}{\rho}} \tag{4.4}$$

Hence, we have the relationships:

$$f_2 = 2f_1 \qquad f_3 = 3f_1 \qquad f_4 = 4f_1 \ldots$$

This is sometimes called the *harmonic series*; the fundamental is called the first harmonic, f_2 is called the second harmonic, f_3 the third harmonic etc. We often refer to the second, third, fourth harmonics etc as the *overtones*.

Note here that the angular frequencies for the continuous string are given by:

$$\omega_n = \frac{n\pi c}{l}$$

so that for the first three harmonics we have:

$$\omega_1 = \frac{3.141c}{l} \qquad \omega_2 = \frac{6.283c}{l} \qquad \omega_3 = \frac{9.425c}{l}$$

These values were compared with lumped parameter approximations in Chapter 2.

Because there is a simple integer relationship between the harmonics and the fundamental, stringed instruments sound melodic when the notes are played together. A drum skin, on the other hand, has the complex relationship $f_2 = 1.5933f_1, f_3 = 2.1355f_1, f_4 = 2.2954f_1, \ldots$ and when these notes are sounded simultaneously, the effect is not nearly as melodic.

PFTA 4.7

Suppose that when the tension in a string is T_1, the fundamental frequency is f_1. Show that the tension T_2 necessary to obtain a fundamental frequency f_2 in the same string is given by:

$$T_2 = \left(\frac{f_2}{f_1}\right)^2 T_1$$

Solution 4.7

If the length of the string is l, then using equation (4.4) it follows that:

$$f_1 = \frac{1}{2l}\sqrt{\frac{T_1}{\rho}} \quad \text{and} \quad f_2 = \frac{1}{2l}\sqrt{\frac{T_2}{\rho}}$$

Hence, we obtain:

$$T_2 = \left(\frac{f_2}{f_1}\right)^2 T_1$$

When listening to a musical sound, it is the pitch of the note that is perceived by the listener. We shall see in Chapter 5 that the pitch of a note is related to its frequency and that it is Mersenne's law which governs the design of stringed musical instruments.

In the piano, each string is associated with a single note. We shall see in Chapter 5 that if we double the frequency, then we obtain an octave increase in the pitch. Consequently, a piano which spans seven octaves requires the highest frequency to be 2^7 times as large as the lowest frequency; hence, using the solution of PFTA 4.7, the ratio of the tensions would be more than 16 000. Clearly, this would lead to an unacceptable variation in tension across the frame. For this reason, all three variables, l, ρ and T in equation (4.3), are varied in the different strings of the piano. In general, short strings are used for the high notes while long strings are used for the low notes. Also, lower notes are produced by strings with higher values of ρ.

Instruments such as the violin, with four strings, and the guitar, with six or twelve strings, have different notes produced by the same string. This is done by holding down the string at a suitable point, hence changing its effective length.

PFTA 4.8

With reference to Mersenne's law, describe the properties of violin or guitar strings which ensure that each string has a different fundamental frequency.

Solution 4.8

The strings on these instruments are of the same length so that differences in fundamental frequencies are obtained by using strings with different densities and different tensions.

4.2 WAVE ENERGY

We shall illustrate the idea of wave energy and its propagation with reference to a travelling wave on a string.

Consider a small element, Δx, of the string. In the motion, this element is stretched to a length Δs, as shown in Figure 4.3. Hence, the work done by the tension is given by:

$$T(\Delta s - \Delta x)$$

The work done is, of course, the change in potential energy, ΔV, hence:

$$\Delta V = T(\Delta s - \Delta x)$$
$$\simeq T[(\Delta x^2 + \Delta y^2)^{1/2} - \Delta x] \qquad \text{since } \Delta s^2 \simeq \Delta x^2 + \Delta y^2$$

Now $\Delta y \simeq (dy/dx) \Delta x$ so that $\Delta V \simeq T\Delta x \, [(1 + (y')^2)^{1/2} - 1]$, where $y' = dy/dx$. If the transverse vibrations are of small amplitude, then $|y'| \ll 1$. Hence, if we expand the term $(1 + (y')^2)^{1/2}$ using the binomial theorem, we have:

$$(1 + (y')^2)^{1/2} \simeq 1 + \tfrac{1}{2} (y')^2$$

and:

$$\Delta V \simeq \tfrac{1}{2}T\Delta x(y')^2$$
$$= \tfrac{1}{2}\rho c^2 \Delta x(y')^2$$

using the solution of PFTA 4.2.

The change in kinetic energy of the element is:

$$\Delta U = \tfrac{1}{2}\rho \Delta x \dot{y}^2$$

since ρ is the mass per unit length of the string.

Hence, the change in total energy is:

$$\Delta E = \Delta U + \Delta V$$
$$\simeq \tfrac{1}{2}\rho(\dot{y}^2 + c^2 y'^2)\Delta x$$

The quantity $D = \tfrac{1}{2}\rho(\dot{y}^2 + c^2 y'^2)$ is the energy per unit length of the string and is called the *energy density*.

PFTA 4.9

Show that the energy density for the harmonic wave:

$$y = A \cos 2\pi(kx - ft)$$

is given by:

$$D = 4\pi^2 \rho A^2 f^2 \sin^2 2\pi(kx - ft)$$

Solution 4.9

$$\dot{y} = 2\pi f A \sin 2\pi(kx - ft)$$
$$y' = -2\pi k A \sin 2\pi(kx - ft)$$

Hence:

$$D = \tfrac{1}{2}\rho 4\pi^2 A^2 \sin^2 2\pi(kx - ft)(f^2 + c^2 k^2)$$

Now, using equation (3.5), $k = f/c$ and it follows that:

$$D = 4\pi^2 \rho A^2 f^2 \sin^2 2\pi(kx - ft)$$

It follows from the solution of PFTA 4.9 that the energy density is itself a harmonic wave whose amplitude is $4\pi^2\rho A^2 f^2$ and whose speed of propagation is $f/k = c$, i.e. energy is propagated with the same speed as the wave.

4.3 FOURIER'S THEOREM

Fourier introduced his famous theorem in the context of heat conduction. However, it is far more wide ranging, occurring in almost all areas of applied mathematics.

Fourier's theorem concerns periodic functions. A function $f(t)$ is said to be *periodic* with period p if for all values of t:

$$f(t + p) = f(t)$$

This property is illustrated in Figure 4.4.

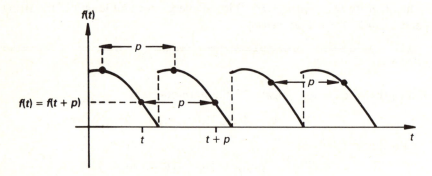

Figure 4.4   The periodic function $f(t)$.

We have already met the periodic trigonometric functions $\sin t$ and $\cos t$ which have period 2π. These functions can be used to produce periodic functions of any period since the functions $\sin (2\pi t/p)$ and $\cos (2\pi t/p)$ have period p. It is convenient in this instance to set $p = 2\tau$ and to assume that the periodic function $f(t)$ is given for $-\tau \leq t \leq \tau$. We shall state Fourier's theorem without proof.

Fourier's theorem tells us that any function of period 2τ can be expressed as a series whose terms are trigonometric functions with period 2τ. Mathematically, we write this as:

$$f(t) = \tfrac{1}{2}a_0 + \sum_{n=1}^{\infty} a_n \cos\left(\frac{n\pi t}{\tau}\right) + \sum_{n=1}^{\infty} b_n \sin\left(\frac{n\pi t}{\tau}\right) \qquad (4.5)$$

The first term is written separately as a matter of convenience. It is equivalent to the case $n = 0$ but the factor $\frac{1}{2}$ is introduced to simplify the algebra later on. The coefficients a_n and b_n are given by:

$$a_n = \frac{1}{\tau} \int_{-\tau}^{\tau} f(t) \cos\left(\frac{n\pi t}{\tau}\right) dt \text{ and } b_n = \frac{1}{\tau} \int_{-\tau}^{\tau} f(t) \sin\left(\frac{n\pi t}{\tau}\right) dt \quad (4.6)$$

PFTA 4.10

Show that if the function in $f(t)$ is an even function, then the coefficients b_n are zero and find an expression for the coefficients a_n. Similarly, if $f(t)$ is an odd function, then the coefficients a_n are zero and we can obtain a similar expression for the coefficients b_n.

Solution 4.10

If $f(t)$ is an even function, then $f(-t) = f(t)$. Now:

$$b_n = \frac{1}{\tau} \int_{-\tau}^{\tau} f(t) \sin\left(\frac{n\pi t}{\tau}\right) dt$$

$$= \frac{1}{\tau} \int_{-\tau}^{0} f(t) \sin\left(\frac{n\pi t}{\tau}\right) dt + \frac{1}{\tau} \int_{0}^{\tau} f(t) \sin\left(\frac{n\pi t}{\tau}\right) dt$$

In the first integral set, $t = -u$ so that $dt = -du$ and hence:

$$\int_{-\tau}^{0} f(t) \sin\left(\frac{n\pi t}{\tau}\right) dt = \int_{\tau}^{0} f(-u) \sin\left(-\frac{n\pi u}{\tau}\right)(-du)$$

Now $f(-u) = f(u)$ and $\sin(-n\pi u/\tau) = -\sin(n\pi u/\tau)$. Also, we can change the order of the limits if we introduce a further negative sign. Hence, the first integral becomes:

$$-\int_{0}^{\tau} f(u) \sin\left(\frac{n\pi u}{\tau}\right) du = -\int_{0}^{\tau} f(t) \sin\left(\frac{n\pi t}{\tau}\right) dt$$

Hence, $b_n = 0$ for all n.
Now:

$$a_n = \frac{1}{\tau} \int_{-\tau}^{0} f(t) \cos\left(\frac{n\pi t}{\tau}\right) \, dt + \frac{1}{\tau} \int_{0}^{\tau} f(t) \cos\left(\frac{n\pi t}{\tau}\right) \, dt$$

Using the same change of variable as before, $t = -u$, in the first integral, we see that:

$$a_n = \frac{1}{\tau} \int_{0}^{\tau} f(u) \cos\left(\frac{n\pi u}{\tau}\right) \, du + \frac{1}{\tau} \int_{0}^{\tau} f(t) \cos\left(\frac{n\pi t}{\tau}\right) \, dt$$

$$= \frac{2}{\tau} \int_{0}^{\tau} f(t) \cos\left(\frac{n\pi t}{\tau}\right) \, dt$$

By a similar process, if $f(t)$ is odd, then $a_n = 0$ for all n and:

$$b_n = \frac{2}{\tau} \int_{0}^{\tau} f(t) \cos\left(\frac{n\pi t}{\tau}\right) \, dt$$

It follows from the solution of PFTA 4.10 that all of the information concerning $f(t)$ may be found from a knowledge of the coefficients a_n and b_n associated with the frequency f_n ($= n/2\tau$). This is the basis of the Fourier analysis method.

We shall not take the mathematics any further. However, it can be shown that as n gets larger so the numerical values of a_n and b_n decrease, and a good approximation can often be obtained using only the first few harmonics, as we shall see in the following example.

Consider the so-called square wave shown in Figure 4.5. Since $f(t)$ is an even function, $b_n = 0$ for all n. Using the solution of PFTA 4.10:

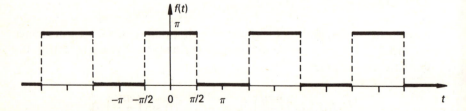

Figure 4.5 A square wave with period 2π.

$$a_n = \frac{2}{\tau} \int_0^\tau f(t) \cos\left(\frac{n\pi t}{\tau}\right) \, dt \qquad \text{where } \tau = \pi$$

$$= \frac{2}{\pi} \int_0^\pi f(t) \cos nt \, dt$$

For $0 \leq t \leq \pi$, we have:

$$f(t) = \begin{cases} \pi & 0 \leq t \leq \pi/2 \\ 0 & \pi/2 < t \leq \pi \end{cases}$$

Hence:

$$a_n = \frac{2}{\pi} \int_0^{\pi/2} \pi \cos nt \, dt$$

$$= 2 \left[\frac{1}{n} \sin nt \right]_0^{\pi/2} \qquad n \neq 0$$

$$= \frac{2}{n} \sin\left(\frac{n\pi}{2}\right)$$

$$a_0 = \frac{2}{\pi} \int_0^{\pi/2} \pi \, dt$$

$$= \pi$$

The values of $|a_n|$ are shown in Figure 4.6.

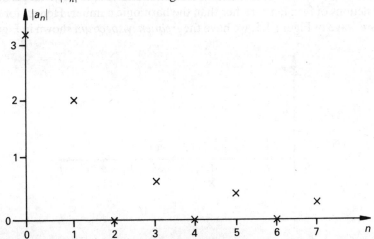

Figure 4.6 The Fourier coefficients a_n for the square wave shown in Figure 4.5.

61

The Fourier series for the square wave is given by:

$$f(t) = \frac{\pi}{2} + \frac{2}{1} \cos t - \frac{2}{3} \cos 3t + \frac{2}{5} \cos 5t - \frac{2}{7} \cos 7t + \ldots$$

If we truncate the series for $f(t)$, we obtain an approximation to $f(t)$. The effect of truncating after various harmonics is shown in Figure 4.7.

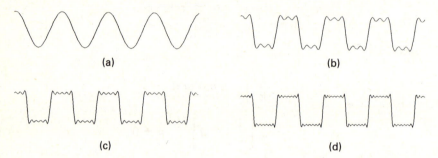

(a)

(b)

(c)

(d)

Figure 4.7 Approximations to $f(t)$ truncating (a) after the first harmonic, (b) after the third harmonic, (c) after the fifth harmonic and (d) after the seventh harmonic.

The process of developing the terms in the Fourier series of a periodic function is often called *Fourier analysis*.

The Fourier coefficients yield the amplitudes associated with each harmonic and the frequency of each harmonic is a multiple of the fundamental frequency f_1 ($= 1/2\tau$). It is usual, in a Fourier analysis, to write amplitudes as functions of frequency rather than the harmonic number. Hence, for the square wave of Figure 4.5, we have the *frequency spectrum* shown in Figure 4.8.

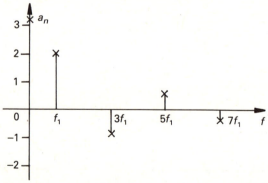

Figure 4.8 Frequency spectrum for the square wave shown in Figure 4.5.

The importance of Fourier's theorem is that it allows us to consider the most complicated waves as a combination of simple waves.

PFTA 4.11

Find the Fourier series and sketch the frequency spectrum for the saw-tooth wave shown in Figure 4.9

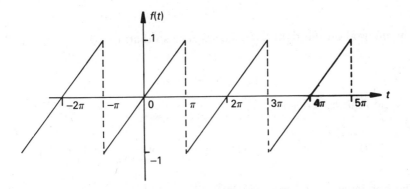

Figure 4.9 The saw-tooth wave.

Solution 4.11

The saw-tooth wave is an odd function of period 2π, hence the Fourier series contains only sine terms and is given by:

$$f(t) = \sum_{n=1}^{\infty} b_n \sin nt \qquad \text{where } b_n = \frac{2}{\pi} \int_0^{\pi} f(t) \sin nt \, dt$$

Now $f(t)$ is a linear function whose graph passes through the points $(-\pi, -1)$ and $(\pi, 1)$. Hence, its equation is given by:

$$f(t) = \frac{t}{\pi} \qquad -\pi \leqslant t \leqslant \pi$$

63

Hence:

$$b_n = \frac{2}{\pi^2} \int_0^\pi t \sin nt \, dt$$

$$= \frac{2}{\pi^2} \left[-\frac{t}{n} \cos nt \right]_0^\pi + \frac{2}{\pi^2 n} \int_0^\pi \cos nt \, dt \text{ (using integration by parts)}$$

$$= \frac{2}{\pi^2} \left(-\frac{\pi}{n} \cos n\pi \right) + \frac{2}{\pi^2 n} \int_0^\pi \cos nt \, dt$$

The integral on the right-hand side is zero so that:

$$b_n = -\frac{2}{n\pi} \cos n\pi$$

and the Fourier series is:

$$f(t) = \frac{2}{\pi} \left(\sin t - \frac{1}{2} \sin 2t + \frac{1}{3} \sin 3t - \frac{1}{4} \sin 4t + \ldots \right)$$

The frequency spectrum is shown in Figure 4.10.

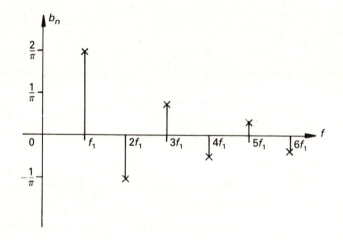

Figure 4.10 Frequency spectrum for the saw-tooth wave.

4.4 LONGITUDINAL WAVES IN SPRINGS

In contrast with waves on strings which are transverse, waves in springs are longitudinal. They are propagated in the same direction as the disturbance of each particle of the spring. A comparison of the two types of wave is shown in Figure 4.11.

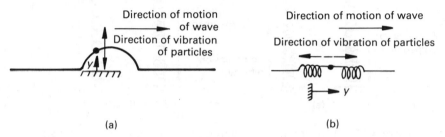

(a)   (b)

Figure 4.11   (a) A transverse wave on a string. (b) A longitudinal wave in a spring.

Hooke's law for a spring says that the extension is proportional to the tension in the spring. The constant of proportionality is the spring's stiffness (see Chapter 2). It is Hooke's law which would enable us to develop the equation describing the propagation of longitudinal waves in springs. However, it is a somewhat more complicated process than the development of transverse waves on strings, and so we shall simply state the result here.

Suppose that the spring has stiffness k, mass per unit length ρ and natural length l. If the disturbance is due to a harmonic wave with angular frequency ω, then the wave form has the equation (cf equation (4.1)):

$$y = \phi(x) \cos(\omega t + \alpha) \tag{4.7}$$

where the function $\phi(x)$ must satisfy suitable conditions at $x = 0$ and $x = h$, the ends of the springs, and also satisfy the differential equation:

$$\frac{d^2\phi}{dx^2} = -\frac{\omega^2\phi}{c^2} \qquad \text{where } c^2 = \frac{kh}{\rho}$$

For fixed ends, we have the boundary conditions $\phi(0) = \phi(h) = 0$ and our mathematical problem is identical to that for the string in Section 4.1. Consequently, we can use all the information from that section, together with Fourier's theorem, to develop solutions for wave propagation in springs.

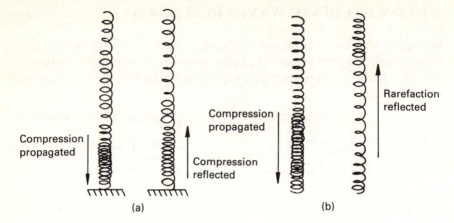

Figure 4.12 Reflection of a longitudinal wave in a Slinkey spring:
(a) a fixed end; (b) a free end.

When developing solutions, we must take into account the difference between the reflection of a wave at a fixed end and the reflection at a free end of a spring. Just as with reflection of a pulse on a string (see Section 3.2), reflection at a fixed end results in a change of phase, whereas reflection at a free end does not. If an end is fixed and a wave is propagated so that it is reflected at that end, then a compression is reflected as a compression together with a change in phase of 180°. At a free end, a compression is reflected as a rarefaction without change in phase. This is easily demonstrated using a Slinkey spring and is illustrated in Figure 4.12.

A similar situation occurs when sound waves are reflected at the end of an organ pipe. In a closed pipe, i.e. a fixed end, a compression is reflected as a compression. In an open pipe, however, we have a free end at which a compression expands rapidly sending a rarefaction back down the pipe.

Longitudinal waves may also be propagated in thin rods. It is usual in such cases to use Hooke's law to relate stress to strain rather than tension to extension. In the following exercise, we develop the parameters for longitudinal waves in a rod.

PFTA 4.12

Suppose that the rod has cross-sectional area A, natural length l and stiffness k. If the tension in the rod is T and its extension is x, then the stress and strain are given by $\sigma = T/A$ and $\varepsilon = x/l$ respectively. Write down Hooke's law in terms of stress and strain. What is the speed of wave propagation in the rod?

Solution 4.12

Hooke's law relates T to x by $T = kx$. Hence:

$$\frac{T}{A} = \frac{kl}{A}\left(\frac{x}{l}\right) \qquad \text{(i.e. } \sigma = E\varepsilon\text{)}$$

where $E = kl/A$ is called *Young's modulus* for the material. Since $c^2 = kl/\rho$, we may write $c^2 = EA/\rho$. Now ρ is the mass per unit length of the rod, hence the density of the material is given by $\rho_0 = \rho/A$. It follows that the wave speed is given by:

$$c = \sqrt{\frac{E}{\rho_0}} \qquad\qquad (4.8)$$

Rods can also transmit transverse waves but we shall not consider these here. The distinction between longitudinal and transverse waves in solid materials is of particular importance in seismology. The material which comprises the inside of the Earth may be forced to vibrate in a transverse or a longitudinal manner and both types of wave may be propagated. When an earthquake occurs, a primary (P) wave, which is longitudinal, and a secondary (S) wave, which is transverse, are propagated into the interior. These waves are reflected and refracted at the interface between different regions of the Earth's interior and are then received at various measuring stations. From the information received, geologists are able to describe the interior structure of the Earth.

5 SOUND WAVES

5.1 PROPAGATION OF SOUND WAVES

Sound waves are mechanical longitudinal waves that can be propagated in solids, liquids or gases. There is a very large range of frequencies over which such mechanical waves can be transmitted. The *audible* range is that range of frequencies which may be detected by the human ear. Mechanical waves in this range are called *sound waves*. The audible range is roughly between 20 Hz and 20 000 Hz. Waves with frequencies lower than the audible range are called *infrasonic* while those with frequencies above the audible range are called *ultrasonic*.

Sound waves are propagated by successive compression and rarefaction of the medium as the wave propagates. We can illustrate this process by considering the motion of a sound pulse down an air-filled tube caused by the oscillations of a piston (see Figure 5.1). If the piston is pushed at one end of the tube, then the air in the vicinity of the piston will be compressed to a pressure above its undisturbed value. The compressed air moves forward compressing the air in front of it and hence sending a *compression pulse* down the tube. As the piston is pulled back, the air in the vicinity of the piston expands and its pressure is reduced to a pressure below its undisturbed value. In this way, a *rarefaction pulse* propagates down the tube. As the piston oscillates, so a continuous train of compression and

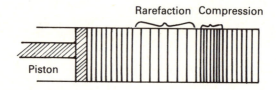

Figure 5.1 Propagation of a compressive pulse in a tube followed by a low pressure pulse called a rarefaction.

rarefaction pulses is propagated down the tube.

The speed of sound propagation in a solid is the speed of longitudinal wave propagation. It is given in terms of the density, ρ_0, and Young's modulus, E, of the material by equation (4.8):

$$c = \sqrt{\frac{E}{\rho_0}}$$

There is a parameter, B, associated with fluids which is analogous to E for solids. It is called the *bulk modulus* and is a measure of the elasticity or compressibility of the fluid. The speed of sound in a fluid is given by:

$$c = \sqrt{\frac{B}{\rho}} \qquad (5.1)$$

where ρ is the fluid density.

For sound propagation in a gas, the changes in pressure occur sufficiently quickly that there is no flow of heat. In such cases, the changes are said to be *adiabatic*. For adiabatic changes, the pressure, p, is related to the density, ρ, by an equation of the form:

$$p = k\rho^\gamma \qquad (5.2)$$

where k and γ are constants depending on the gas. γ is the ratio of the two specific heats for the gas. In this case, the bulk modulus is given by:

$$B = \gamma p_0 \qquad (5.3)$$

where p_0 is the undisturbed pressure. Hence, the velocity of sound is given by:

$$c = \sqrt{\frac{\gamma p_0}{\rho_0}} \qquad (5.4)$$

5.2 SOUND WAVES IN PIPES

The propagation of sound in a pipe is the basic mechanism for wind instruments. A disturbance is created at one end of the pipe, e.g. in an organ pipe, a jet of air strikes a sharp edge; in a flute, the player blows across a hole; and in a clarinet, a reed vibrates when blown. This disturbance then propagates as a wave down the pipe.

69

PFTA 5.1

Make a list of other wind instruments and state how the disturbance is created at the end of the pipe. Can you think of any other circumstances where waves are propagated in a pipe?

Solution 5.1

In the recorder, air strikes a sharp edge in a similar manner to the organ pipe. The bassoon, oboe and saxophone are wind instruments which incorporate vibrating reeds. The piccolo is a small instrument of the flute family in which air is blown over a hole. Brass instruments, such as the horn, trombone, trumpet and tuba, are activated by vibrating the performer's lips which play the same role as the reed does in a clarinet.

There are many situations, other than musical instruments, in which sound waves are propagated in a pipe. For example, the 'water hammer' is that very unpleasant sound which sometimes occurs in plumbing when a tap is turned on. It is the compressibility of the water that allows the weird noise to travel down the pipe-work. The noise from a motor vehicle comes from sound propagated in its exhaust pipe.

To understand wave propagation in wind instruments, we shall consider the simple case of a straight, uniform tube such as an organ pipe. Longitudinal waves travel along the pipe and are reflected at the other end.

We have already noted in Section 4.4 that the reflection of sound waves at the end of a pipe is very similar to the reflection of longitudinal waves at the end of a spring. A change in phase must occur at a closed end.

Interference with the reflected wave sets up a standing wave pattern, the nature of which depends on the conditions at the end of the pipe. If it is a closed end, then it will be a node and the standing wave will always have an odd number of quarter-wavelengths as shown in Figure 5.2(a). If, on the other hand, the end is open, then the standing wave will have an even number of quarter-wavelengths as shown in Figure 5.2(b).

If we consider the closed pipe, we see that, in the fundamental mode, one-quarter of the wavelength is equal to the length of the pipe, i.e. $\lambda_1/4 = l$ where l is the length of the pipe. If the speed of the wave is c, then the fundamental frequency is related to the fundamental wavelength by the expression $c = f_1\lambda_1$. Hence, the fundamental frequency is given by $f_1 = c/4l$.

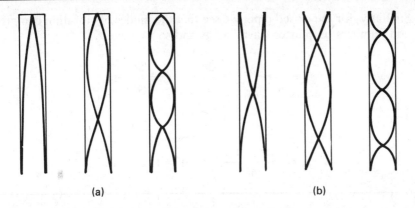

(a)   (b)

Figure 5.2 Standing waves in a pipe: (a) a closed pipe; (b) an open pipe.

PFTA 5.2

Show that the natural frequencies of the organ pipe are given by the following formulae:

$$f_n = \frac{nc}{2l} \qquad n = 1, 2, 3, \ldots \text{ for an open pipe}$$

$$f_n = \frac{(2n-1)c}{4l} \qquad n = 1, 2, 3, \ldots \text{ for a closed pipe}$$

Solution 5.2

If we consider the diagrams in Figure 5.2(b), it is easy to see that in the fundamental mode for the open pipe there is one half-wavelength, so that $\lambda_1 = 2l$ and hence $f_1 = c/2l$. Similarly, in the second, third, fourth harmonics, we have two, three, four half-wavelengths etc. Consequently:

$$f_1 = \frac{c}{2l} \qquad f_2 = \frac{2c}{2l} \qquad f_3 = \frac{3c}{2l} \qquad f_4 = \frac{4c}{2l} \text{ etc}$$

that is:

$$f_n = \frac{nc}{2l} \qquad n = 1, 2, 3, \ldots$$

71

Similarly, for the closed pipe, we see that the modes of vibration comprise odd numbers of quarter-wavelengths and so we have:

$$f_1 = \frac{c}{4l} \qquad f_2 = \frac{3c}{l} \qquad f_3 = \frac{5c}{4l} \quad \text{etc}$$

That is:

$$f_n = \frac{(2n - 1)c}{4l} \qquad n = 1, 2, 3, \ldots$$

The importance of the solution of PFTA 5.2 is that the quality of sound from a stopped pipe is different from that of the open pipe, since in the stopped pipe only odd harmonics are present. The presence of the even overtones in the open pipe allows a much richer sound. It is the combination of fundamental and overtones which provides the characteristics of different musical instruments.

5.3 SOUND AND MUSIC

Sound waves may be generated in any medium by producing a suitable disturbance. Vibrations in which a pattern is repeated at regular intervals produce sound waves that are pleasing to listen to. Vibrations that are not periodic produce noises that are not pleasant to the ear. Musical instruments provide a mechanism for producing periodic sound waves in air.

We have already discussed transverse standing waves on strings in Chapter 4, so we shall use some of the results to explain why the vibrations of a violin string produce a different 'sound' to that from a piano string.

We know from Mersenne's law, equation (4.3), that the natural frequencies are, using the notation of Chapter 4, given by:

$$f_n = \frac{n}{2l} \sqrt{\frac{T}{\rho}} \tag{5.5}$$

Why, then, do these two instruments sound different? The answer lies in the fact that, with the exception of the tuning fork, no instrument produces a pure tone; all notes include overtones. We can, by choosing suitable values for l, T and ρ, produce a note with frequency given by Mersenne's law. However, the mechanisms by which the note is produced, bowing the violin string or striking the piano wire, mean that it is impossible to produce the pure note – higher harmonics will always be present as well.

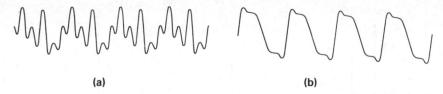

(a)   (b)

Figure 5.3   Waveform for a note of frequency 440 Hz, concert A, on
(a) violin and (b) piano.

This isn't the whole story – the sounding box of the instrument also plays a
crucial role in the sound production. We shall say a little more about this
later. In Figure 5.3 we see the waveform for a note of frequency 440 Hz, a
concert A, played on a violin and on a piano. Notice that there are loud
higher harmonics, especially the second and fifth, in the violin. It is the
presence of these overtones which provides the quality of the sound,
usually called the *timbre*.

Let us now return to the effect of the sounding box on a stringed
instrument. If we pluck a taught string, very little sound is emitted. We
require what is known as a *resonator* to provide a sufficiently large disturbance to be noticed by the ear.

A simple demonstration is provided using a tuning fork. If the fork is
struck, a note is emitted. If the fork is then placed so that its base rests on a
hollow box, the resonator, then there is a considerable increase in the
volume of the note.

The action of the resonator is to take the vibration and to amplify it. This
is exactly what happens in the body of a violin. The strings provide the
vibration and the body acts as an amplifier. What we hear comes from the
disturbance in the air produced by the resonator, not the strings. The ratio
of the output amplitude from the resonator to the input amplitude from the
strings is called the *gain* of the amplifier; however, the gain varies with
frequency. Consequently, the resulting waveform we hear is not the same
as that of the original vibration from the strings. We say that the body of
the violin has a *formant* characteristic, and it is this characteristic which
alters the frequency spectrum of whichever note is being played. In general, it is the formant characteristic which tells us that we are listening to a
violin or to a piano. Indeed, every violin has its own formant characteristic
and these characteristics are sufficiently different that a suitably trained
musical ear can distinguish between different violins.

We know from Fourier's theorem that a general periodic function can be
considered to be a combination of simple sinusoidal waves, i.e. a sum of
the possible harmonics. The frequency spectrum allows us to determine the
contribution that each overtone makes to the sound. In Figure 5.4, we see
the frequency spectrum for a concert A played on the violin compared with
that played on the piano.

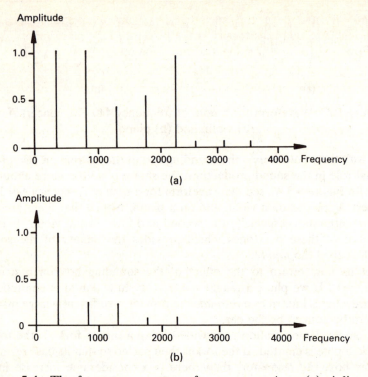

Figure 5.4 The frequency spectrum for a concert A on (a) violin and (b) piano.

A second property of a sound wave which is important from the listener's point of view is its *loudness*. This is a subjective attribute which depends on the listener. However, the loudness is related to an objective attribute, the *intensity* of the sound, which is proportional to the power of the source.

The Weber–Fechner law is an empirical relationship which relates subjective perceptions, i.e. those qualities that depend on the individual, and objective quantities, i.e. those qualities that are independent of the individual and can be measured uniquely. If S is a measure of the subjective entity and U is the measure of the objective entity, then the Weber–Fechner law gives:

$$S = k \log U \tag{5.6}$$

where k is a constant. If we denote the loudness of sound by L and its intensity by I, then the Weber–Fechner law yields:

$$L = 10 \log_{10} I \tag{5.7}$$

where it is usual to use units such that $k = 10$ and to take logarithms to the base ten. The unit of intensity is the picowatt per square metre and the corresponding unit of loudness is called the decibel (dB).

PFTA 5.3

Suppose that I_0 is the intensity at which the lowest perceivable loudness, L_0, occurs. Show that at loudness L:

$$L - L_0 = 10 \log_{10} \left(\frac{I}{I_0} \right)$$

and that a doubling of intensity leads to an increase in loudness of about 3 dB

Solution 5.3

Using equation (5.7):

$$L - L_0 = 10 \log_{10} I - 10 \log_{10} I_0$$

$$= 10 \log_{10} \left(\frac{I}{I_0} \right)$$

Suppose that the loudness is increased to $L + \Delta L$ with a corresponding intensity, I_1, then:

$$L + \Delta L - L_0 = 10 \log_{10} \left(\frac{I_1}{I_0} \right)$$

$$\Delta L = 10 \log_{10} \left(\frac{I_1}{I_0} \right) - 10 \log_{10} \left(\frac{I}{I_0} \right)$$

$$= 10 \log_{10} \left(\frac{I_1}{I} \right)$$

It follows, then, that if the intensity is doubled so that $I_1 = 2I$, then:

$$\Delta L = 10 \log_{10} 2$$

$$\simeq 3 \text{ dB}$$

The range of loudness values which the human ear can detect is very large indeed and is shown by the examples in the following exercise.

PFTA 5.4

Suppose that we define the threshold of hearing, L_0, to be 0 dB, so that the corresponding intensity is 1 pWm^{-2}. The following sound intensities are produced by certain well-known phenomena:

10 pW,   100 pW,   3200 pW, 10^6 pW,   2.5 × 10^7 pW,   3.8 × 10^8 pW,
3 × 10^{10} pW, 10^{11} pW, 10^{12} pW

Make a table relating the intensities to the corresponding loudness values.

Solution 5.4

Phenomenon	Loudness (dB)	Intensity (pW)
Threshold of hearing	0	1
Leaves rustling	10	10
Whisper	20	100
Library	35	3200
Busy office	60	10^6
Vacuum cleaner	74	2.5 × 10^7
Heavy traffic	86	3.8 × 10^8
Loud rock music	105	3 × 10^{10}
Jet aircraft	110	10^{11}
Threshold of pain	120	10^{12}

The third important attribute of sound in a musical context is the *pitch*, which is related to the frequency of the sound. The pitch of a sound wave is analogous to the colour of a light wave in the visible spectrum. It is the colour that the eye perceives and hence determines the frequency for the viewer.

The pitch, p, like the loudness, is a subjective property and according to the Weber-Fechner law is related to the frequency, f, by the equation:

$$p = k \log f \tag{5.8}$$

Just as with the perception of loudness, the human ear is sensitive to a large range of frequencies from about 20 Hz at the lower end up to about 20 000 Hz at the upper end. These values, of course, vary from one individual to another and the upper value tends to decrease with age. It has also been suggested that sensitivity to the higher frequencies is being severely eroded among young people who are exposed to an excessive amount of loud music from disco equipment or personal stereo systems.

The logarithmic relation between pitch and frequency given by equation (5.8) means that the range of values of p is about $3k$.

In the Western scale of music, notes are grouped together in octaves where the frequency of the highest note in the octave is double that of the lowest note. Also, on the equal-tempered scale, the octave is subdivided into 12 semitones, each of which is subdivided into 100 equal units called cents. For this reason, we choose the value of k to be given by:

$$k = \frac{1200}{\log 2} \tag{5.9}$$

PFTA 5.5

Derive the expression for k given by equation (5.9).

Solution 5.5

Suppose that a pitch, p, corresponds to the frequency, f, then from equation (5.8):

$$p = k \log f$$

The pitch p_1, which is one octave higher than p, differs from p by 12 semitones, in each of which there are 100 cents. Hence:

$$p_1 = p + 12 \times 100$$

Since this pitch corresponds to a doubling of the frequency:

$$p_1 = k \log (2f)$$

77

Hence:

$$k \log (2f) = k \log f + 1200$$
$$k \log 2 + k \log f = k \log f + 1200$$
$$k = \frac{1200}{\log 2}$$

The equal-tempered scale can be described by considering the notes on a piano. Each white key has a name given by a letter A, B, C, D, E, F or G (see Figure 5.5). If we include the black keys, then the pitch increment between successive keys is one semitone.

Figure 5.5   A section of a piano keyboard.

The sequence of notes CDEFGABC forms an octave of the scale of C major. Notice that the interval between the notes C and D is two semi-tones, i.e. one full tone, since the white keys corresponding to C and D are separated by a black key.

PFTA 5.6

What are the pitch increments between the notes C and D, D and E, E and F, F and G, G and A, A and B, B and C?

Solution 5.6

C and D: one full tone
D and E: one full tone
E and F: one semitone
F and G: one full tone
G and A: one full tone
A and B: one full tone
B and C: one semitone

All major scales have this structure: tone, tone, semitone, tone, tone, tone, semitone (.T.T.S.T.T.T.S.).

If we choose the note C nearest to the middle of the piano, then we have the note called middle C. The A above middle C is the concert A with a frequency of 440 Hz.

PFTA 5.7

Determine the frequencies of all of the notes in the octave of the scale of C major starting at middle C. Give the values correct to four significant figures.

Solution 5.7

We know that the frequency is doubled across the octave and that the octave is divided into 12 equal-value semitones. If follows therefore that an increase in pitch of one semitone produces an increase in frequency by a factor of $2^{1/12}$, i.e. a factor of approximately 1.0595. If f is the frequency of middle C, then, since concert A is nine semitones above middle C and has a frequency of 440 Hz:

$$440 = (2^{1/12})^9 f$$

Hence, $f = 261.6$ Hz. Using the solution of PFTA 5.6, we can now develop the frequencies in the key of C major starting at middle C as follows:

C – 261.6 Hz
D – 293.7 Hz
E – 329.6 Hz
F – 349.2 Hz
G – 392.0 Hz
A – 440.0 Hz
B – 493.9 Hz
C – 523.3 Hz

The black notes on the piano are one semitone higher than the corresponding white notes to their left. Similarly, they are one semitone lower than the corresponding white notes to their right. So, for example, the note between C and D is called either C sharp, written $C^\#$, or D flat, written $D^\flat$.

In the theory of music, there is a distinction between the two notes but we will not pursue this here. With the sharps or flats added, we now have names for all of the keys. If we start with any key, say for example G, and progress upwards in the order (.T.T.S.T.T.T.S.), then we have an octave in the scale of G major given by GABCDEF$^\#$G.

PFTA 5.8

Write down the notes in the major scales of D, A and F. Note that in the scale of F major, it is usual to write B$^\flat$ rather than A$^\#$.

Solution 5.8

D major: DEF$^\#$GABC$^\#$D

A major: ADC$^\#$DEF$^\#$G$^\#$A

F major: FGAB$^\flat$CDEF

We finish this section by mentioning the so-called minor key in which the interval structure in the octave is of the form (.T.S.T.T.S.T+S.S.). The scale of C minor would therefore be CDE$^\flat$FGA$^\flat$BC.

PFTA 5.9

Write down the notes in the minor scale of A.

Solution 5.9

A minor: ABCDEFG$^\#$A

In this equal temperament, a scale in either of the two key modes may start on any note and progress upwards through an octave using either sharps or flats if necessary. Hence, we have 12 keys in each of the major and minor modes.

PFTA 5.10

If you have had a musical training, you will be familiar with the notation for writing music in terms of notes on a stave. The examples shown in Figure 5.6 give the scales of B♭ major and A minor respectively. Write down the frequencies of the notes in each scale. Assume that the B♭ major scale starts at the B♭ above middle C and the A minor scale starts at the A below middle C.

(a) (b)

Figure 5.6 One octave of each of the scales (a) B♭ major and (b) A minor.

Solution 5.10

The frequencies of the notes in the scale of B♭ major starting at the B♭ above middle C and in the scale of A minor starting at the A below middle C are as follows:

B♭ major: B♭ – 466.2 Hz   A minor: A – 220.0 Hz
           C – 523.3 Hz   B – 246.9 Hz
           D – 587.4 Hz   C – 261.6 Hz
           E♭ – 622.3 Hz   D – 293.7 Hz
           F – 698.5 Hz   E – 329.6 Hz
           G – 784.1 Hz   F – 349.2 Hz
           A – 880.1 Hz   G# – 415.3 Hz
           B♭ – 932.4 Hz   A – 440.0 Hz

6 GENERAL PROPERTIES OF WAVES

6.1 DOPPLER EFFECT

We are all familiar with the characteristic change in the pitch of the sound of the siren of an emergency vehicle as it passes. This phenomenon is called the *Doppler effect*. We can explain the effect using the relationship between wave velocity and frequency given by equation (3.5). We assume that the velocity, c, of the wave is independent of the velocity of the source of the wave.

Suppose, in the first instance, that the wave source is stationary at the point P and that a stationary observer receives the waves at Q. If the distance PQ is d and it takes time t for the signal to travel from P to Q, then $d = ct$. Hence, using equation (3.5):

$$d = f\lambda t$$

where f is the frequency of the source.

Now suppose that the source is moving towards Q with speed u. Then, in time t, P will move to P_1 and the waves emitted from the source will be 'squeezed' into the distance P_1Q, (see Figure 6.1). Now in this time t the same number of waves is emitted by the source, hence the observer perceives that the wave has a wavelength λ_1, where:

$$ut + f\lambda_1 t = d$$

as shown in Figure 6.1. It follows, then, that

$$ut + f\lambda_1 t = f\lambda t$$

and consequently:

$$\lambda_1 = \lambda - \frac{u}{f} \tag{6.1}$$

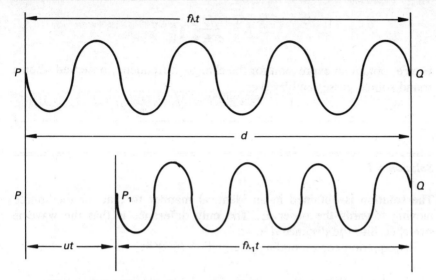

Figure 6.1 Change in wavelength due to a moving source.

If the observer perceives the wave to have a frequency f_1, then, since the speed of the wave is independent of the speed of the source, we have:

$$c = f\lambda = f_1\lambda_1$$

so that, using equation (6.1):

$$\frac{c}{f_1} = \frac{c}{f} - \frac{u}{f} \qquad (6.2)$$

hence:

$$f_1 = \frac{fc}{c - u} \qquad (6.3)$$

Equation (6.3) shows how the frequency of the wave depends on the velocity of the source; as the velocity increases, so the frequency perceived by the observer also increases.

___

PFTA 6.1

Show that if the source is moving away from the observer with speed v, then the frequency perceived by the observer is given by:

$$f_2 = \frac{fc}{c + v}$$

Hence, obtain an expression for the change in frequency observed when a sound source passes an observer.

Solution 6.1

The solution is obtained in an identical manner to that for the source moving towards the observer. The only difference is that the wave is stretched into the distance $d + vt$.
The equation corresponding to equation (6.2) is:

$$\frac{c}{f_2} = \frac{c}{f} + \frac{v}{f}$$

and hence:

$$f_2 = \frac{fc}{c + v}$$

If a source of sound, such as a siren on an emergency vehicle, approaches and passes an observer with speed u, it follows that the frequency is given by equation (6.3) as:

$$f_1 = \frac{fc}{c - u}$$

As it moves away, the frequency is given by f_2. Hence, the change in frequency is given by:

$$f_1 - f_2 = \frac{fc}{c - u} - \frac{fc}{c + u}$$

$$= \frac{2fcu}{c^2 - u^2}$$

Suppose now that the source is at rest and the observer moves away from the source with speed u. In this case, we can superimpose a velocity $-u$ on

84

the whole system of observer, source and wave. This brings the observer to rest, it gives the source a velocity $-u$, and the wave velocity is $c - u$. Hence, we have the situation of a source moving towards the observer and the perceived frequency is given by:

$$f_3 = \frac{f(c - u)}{(c - u) - (-u)}$$

$$= \frac{f(c - u)}{c}$$

PFTA 6.2

What is the frequency perceived by an observer moving with velocity u away from a source which is itself moving towards the observer with velocity v?

Solution 6.2

The situation is exactly the same as that for a source moving towards the observer with velocity $v - u$. Equation (6.3) gives the frequency perceived by the observer as $fc/[c - (v - u)]$.

6.2 BEATS

Consider the superposition of two sinusoidal waves with frequencies f_1 and f_2 and with the same amplitude A. If the wave profiles are y_1 and y_2, then from equation (3.6):

$$y_1 = A \cos 2\pi(k_1 x - f_1 t)$$
$$y_2 = A \cos 2\pi(k_2 x - f_2 t)$$

Hence, the superposed wave profile is:

$$y = y_2 + y_1$$
$$= 2A \cos 2\pi \left[\frac{(k_2 + k_1)}{2} x + \frac{(f_2 + f_1)}{2} t \right] \cos 2\pi \left[\frac{(k_2 - k_1)}{2} x - \frac{(f_2 - f_1)}{2} t \right]$$

It is convenient to use the notation:

$$f_1 = f, \quad f_2 = f + \Delta f, \quad k_1 = k \quad \text{and} \quad k_2 = k + \Delta k$$

Then:

$$y = 2A \cos 2\pi \left[\left(k + \frac{\Delta k}{2} \right) x - \left(f + \frac{\Delta f}{2} \right) t \right] \cos 2\pi \left(\frac{\Delta k}{2} x - \frac{\Delta f}{2} t \right) \quad (6.4)$$

This resulting wave profile is interesting because the first cosine term represents a wave of frequency $f + \Delta f/2$, which is the average frequency of the two constituent waves. The second cosine term, however, has a frequency $\Delta f/2$ and represents a wave which changes far more slowly than the constituent waves. The wave profile given by equation (6.4) is shown in Figure 6.2.

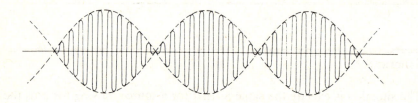

Figure 6.2 The wave profile given by equation (6.4), $t = 0$.

We can think of the wave as having the sinusoidal component $\cos 2\pi$ $[(k + \Delta k/2)x - (f + \Delta f/2)t]$ with a variable amplitude $2A \cos 2\pi[(\Delta k/2)$ $x - (\Delta f/2)t]$. In the case when the two frequencies are nearly equal, so that Δf is small, the sinusoidal components of the two constituent waves and the superimposed wave are nearly equal. However, the superimposed wave has an amplitude which oscillates with the low frequency $\Delta f/2$. The effect is to produce *beats* in which the varying amplitude gradually builds up to a maximum and then slowly dies down, this pattern being repeated with period $2/\Delta f$.

Beats may be heard when two notes of nearly equal frequency are sounded simultaneously. The ear will hear the notes with a slowly varying intensity so that the loudness of the notes increases and decreases with a frequency equal to one-half the difference between the frequencies of the two notes.

PFTA 6.3

If two neighbouring piano keys are sounded together, show that the resulting beat frequency is approximately $0.0297f$, where f is the frequency of the lower note.

The lowest note on a piano is a low A with a frequency 27.5 Hz. The highest note is a high C with a frequency 4186 Hz. Calculate the beat frequency when (a) the two lowest notes and (b) the two highest notes are sounded simultaneously.

Solution 6.3

Neighbouring keys on the piano differ by one semitone. Hence, using the solution of PFTA 5.7, if the lower note has a frequency f, the next note one semitone higher has a frequency $2^{1/12}f$. Now, the beat frequency is given by one-half the difference in frequencies, so that in this case it would be:

$$\frac{2^{1/12}f - f}{2} \simeq 0.0297f$$

(a) The beat frequency when the two lowest notes, A and B^b, are sounded together is approximately 0.82 Hz.
(b) The second highest note is B with a frequency $2^{-1/12}f$, where $f = 4186$ Hz, the frequency of the high C. Hence, the beat frequency when the two notes are sounded together is given by:

$$\frac{f - 2^{-1/12}f}{2} \simeq 0.0281f$$

$$= 117.5 \text{ Hz}$$

The following segment of pseudo-code provides a procedure to illustrate the beats developed by the superposition of the two profiles:

$$y_1 = A \cos 2\pi kx \quad \text{and} \quad y_2 = A \cos 2\pi(k + \Delta k)x$$

The superposed profile is:

$$y = 2A \cos 2\pi[(k + \Delta k/2)x] \cos 2\pi[(\Delta k/2)x]$$

$$= y_{\text{env}} \cos 2\pi[(k + \Delta k/2)x]$$

The profile y is contained between the two envelope curves $\pm y_{env}$.

```
PROCEDURE beats (A, k, delta_k: REAL)
CONSTANT
   two_pi := 6.283185308;
   number_of_points := 200;
VAR
   i: INTEGER;
   x[1..201], y[1..201], y_envelope[1..201]: ARRAY OF REAL;
BEGIN
   DO i = 1, number_of_points+1
      x[i] := FLOAT(i−1)/FLOAT(number_of_points/2);
      y_envelope[i] := 2.0*A*COS(two_pi*delta_k*x[i]);
      y[i] := y_envelope[i]*COS(two-pi*(k+delta_k/2.0)*x[i]);
   END DO;
   set_polyline(number_of_points+1, x, y_envelope)
   set_polyline(number_of_points+1, x, −y_envelope)
   set_polyline(number_of_points+1, x, y)
END;
```

Choose the graphics window to be defined by the rectangle $(0, -1)$, $(2, -1)$, $(2, 1)$ and $(0, 1)$. Suitable values of the parameters are $A = 1.0$, $k = 20.0$ and delta_k $= 2.0$.

6.3 DISPERSION

So far, we have considered waves for which the frequency is independent of the velocity. There are, however, situations in which different frequencies are propagated at different velocities. Such a phenomenon is called *dispersion*. It occurs, for example, in transverse vibrations in rods and in the waves on the surface of a liquid. A medium in which waves exhibit dispersion is called a *dispersive medium*.

We have seen, in Section 4.2, that for a single wave the energy is propagated with the same speed as the wave itself, and this is the speed with which the maximum amplitude is propagated. Now consider what happens when a number of waves are superposed to form a *group*. We shall consider only two waves since the superposition of three or more waves yields the same result. Equation (6.4) gives the superposed wave profile:

$$y = 2A \cos 2\pi \left[\left(k + \frac{\Delta k}{2} \right) x - \left(f + \frac{\Delta f}{2} \right) t \right] \cos 2\pi \left(\frac{\Delta k}{2} x - \frac{\Delta f}{2} t \right)$$

which is illustrated in Figure 6.2.

For this wave group, the maximum amplitude propagates with the velocity of the envelope, the dashed curve in Figure 6.2. This velocity is called the *group velocity*, c_g, of the wave. From equation (6.4), we see that:

$$c_g = \frac{\dfrac{\Delta f}{2}}{\dfrac{\Delta k}{2}}$$

$$= \frac{\Delta f}{\Delta k}$$

Since $k = 1/\lambda$, we have:

$$\Delta k \simeq \frac{dk}{d\lambda}\Delta\lambda$$

$$= -\frac{1}{\lambda^2}\Delta\lambda$$

and:

$$c_g \simeq -\lambda^2 \frac{\Delta f}{\Delta\lambda}$$

Hence, in the limit as $\Delta\lambda \to 0$, we have:

$$c_g = -\lambda^2 \frac{df}{d\lambda}$$

Now $f = c/\lambda$ so that:

$$c_g = -\lambda^2 \left(-\frac{c}{\lambda^2} + \frac{1}{\lambda}\frac{dc}{d\lambda} \right)$$

$$= c - \lambda \frac{dc}{d\lambda} \tag{6.5}$$

In a non-dispersive medium, the velocity, c, is independent of the wavelength, λ, so that $dc/d\lambda = 0$ and hence $c_g = c$. In this case, the wave and the envelope propagate together. However, in a dispersive medium, c is a function of λ and the group velocity is given by equation (6.5). In this case, the envelope moves at a different speed compared with that of the wave and the wave itself appears to move through the wave envelope.

A simple experiment to illustrate group velocity using an overhead projector is described in the article by Jacobs, listed in the bibliography.

PFTA 6.4

For waves on the surface of a deep liquid, the relationship between the velocity, c, and the wavelength, λ, is given by:

$$c^2 = \frac{g\lambda}{2\pi} + \frac{2\pi T}{\lambda\rho}$$

where T is the surface tension, ρ is the density and g is the acceleration due to gravity. Find an expression for the group velocity of the waves.

Solution 6.4

Differentiating c^2 with respect to λ, we obtain:

$$2c\frac{dc}{d\lambda} = \frac{g}{2\pi} - \frac{2\pi T}{\lambda^2\rho}$$

so that:

$$\frac{dc}{d\lambda} = \frac{g}{2c2\pi} - \frac{\pi T}{\lambda^2\rho c}$$

Hence, using equation (6.5):

$$c_g = c - \lambda\frac{dc}{d\lambda}$$

$$= c - \frac{\lambda g}{4\pi c} + \frac{\pi T}{\lambda\rho c}$$

$$= \frac{1}{c}\left(c^2 - \frac{\lambda g}{4\pi} + \frac{\pi T}{\lambda\rho}\right)$$

$$= \frac{g\lambda}{4\pi c} + \frac{3\pi T}{\lambda\rho}$$

6.4 MODULATION

When radio waves are used to transmit a signal, the required signal is not transmitted directly. It is modified in a manner known as *modulation*. We shall examine the situation in which a signal is transmitted using *amplitude modulation*.

When no information is being transmitted, we have a pure tone represented by the harmonic wave:

$$y = A \cos 2\pi(kx - ft)$$

This wave is called the *carrier wave*. In order to transmit information, the amplitude itself is allowed to vary as a function of time. To illustrate the point, suppose we wish to transmit the signal:

$$A = a + b \cos 2\pi pt$$

from the point at which $x = 0$. Since this must travel as a wave with velocity c, it must, at the general point x, be a function of $x - ct$. Now, since $x - ct = x - ft/k$, a function of $x - ct$ can be written as a function of $kx/f - t$. Hence, we take:

$$A = a + b \cos 2\pi p\left(\frac{kx}{f} - t\right) \qquad (6.6)$$

PFTA 6.5

Show that the wave:

$$y = A \cos 2\pi(kx - ft)$$

with A given by equation (6.6) may be written as the superposition of three waves, one with amplitude a and frequency f and the other two with the same amplitude $b/2$ and frequencies $f \pm p$.

Solution 6.5

$$y = \left[a + b \cos 2\pi p\left(\frac{kx}{f} - t\right)\right] \cos 2\pi(kx - ft)$$

$$= a \cos 2\pi(kx - ft) + b \cos 2\pi p\left(\frac{kx}{f} - t\right) \cos 2\pi(kx - ft)$$

The second term on the right-hand side is of the form $b \cos \alpha \cos \beta$, which may be written as:

$$\frac{b}{2} \cos (\beta + \alpha) + \frac{b}{2} \cos (\beta - \alpha)$$

With $\alpha = 2\pi p(kx/f - t)$ and $\beta = 2\pi(kx - ft)$, we have:

$$\beta + \alpha = 2\pi(f + p) \left(\frac{kx}{f} - t \right)$$

$$\beta - \alpha = 2\pi(f - p) \left(\frac{kx}{f} - t \right)$$

Hence:

$$y = a \cos 2\pi(kx - ft) + \frac{b}{2} \cos \left[2\pi(f + p) \left(\frac{kx}{f} - t \right) \right]$$

$$+ \frac{b}{2} \cos \left[2\pi(f - p) \left(\frac{kx}{f} - t \right) \right]$$

That is, we have the superposition of three waves: one wave has amplitude a and frequency f, the other two waves have the same amplitude $b/2$ with frequencies $f \pm p$.

The solution of PFTA 6.5 shows us how the original signal can be recovered from the modulated wave. The receiver will observe three waves with frequencies f and $f \pm p$ and corresponding amplitudes a and $b/2$. Hence, the original signal $a + b \cos 2\pi pt$ can be reconstructed.

The question is: Why do we go to such lengths to transmit this signal? The answer lies in the fact that in radio transmission the frequency p will be in the audio range. However, to transmit a signal in this range would require an aerial of totally unmanageable length. The frequency of the carrier wave is chosen to be high so that the three frequencies f and $f \pm p$ are sufficiently high to require an aerial of manageable size.

Amplitude modulation (AM) is used on the AM wave band for radio transmissions. High quality radio broadcasts are transmitted using frequency modulation, (FM), which is described mathematically in a similar manner to AM.

6.5 INTERFERENCE

The physical effect of the superposition of two or more waves is called *interference*. In the case considered in Section 3.3, we saw that if waves of the same frequency and nearly equal phase are added, the amplitude is increased by the superposition. Such waves are said to interfere *constructively*. Suppose now that the waves have the same amplitude, $A_1 = A_2$, but are almost phase opposed, i.e. $\phi \simeq \pi$, then using equation (3.8):

$$A = (A_1^2 + A_2^2 - 2A_1A_2)^{1/2} = 0 \quad \text{and} \quad \tan \psi = 0 \quad (\text{i.e. } \psi = 0)$$

This means that the two waves cancel each other out. Such waves are said to interfere *destructively*.

Constructive and destructive interference can be seen in many different physical phenomena, e.g. the rainbow pattern seen in a thin oil film by the roadside is due to interference of light waves; and the apparent wave motion noticeable in the railings on motorway bridges is an interference phenomenon which is similar to that which generates the patterns which you can see in the folds of net curtains. If you have access to a personal computer with a graphics facility, you might like to generate similar patterns, known as *Moiré fringes*, as shown in Figure 6.3. The following procedure in pseudo-code will develop the Moiré patterns shown in Figure 6.3.

Figure 6.3 Moiré patterns developed by drawing sets of concentric circles whose centres are a small distance apart.

93

```
PROCEDURE moire (max_radius, x_centre, y_centre, x_inc: REAL)
CONSTANT
   number_of_circles := 40;
   number_of_points := 200;
   two_pi := 6.283185308;
VAR
   count, c, i: INTEGER;
   max_radius, x_centre, y_centre, x_inc, theta, rad: REAL;
   x[1..201], y[1..201]: ARRAY of REAL;
BEGIN
   count := 0;
   DO WHILE (count.LT.2)
   DO i = 1, number_of_circles
      radius := FLOAT(i)*max_radius/FLOAT(number_of_circles);
      DO j = 1, number_of_points+1
         theta := FLOAT(j-1)*two_pi/FLOAT(number_of_points);
         x[i] := x_centre-x_inc+radius*COS(theta);
         y[i] := y_centre+radius*SIN(theta);
      END DO;                          {j = 1,...}
      set_polyline(number_of_points+1, x, y)
   END DO;                             {i = 1,...}
   count := count+1;
   x_centre := x_centre+2.0*x_inc
   END DO;                             {do while}
END;
```

If the graphics window is chosen to be the unit square with corners at
(0, 0), (1, 0), (1, 1) and (0, 1), then choose x_centre = 0.5,
y_centre = 0.5, x_inc = 0.05 and max_radius = 0.45.

BIBLIOGRAPHY

Here is a short list of textbooks and articles which the interested reader can use to develop a deeper understanding of wave motion. Those texts which develop the mathematics of waves require a knowledge of mathematical techniques usually covered in a first undergraduate course for scientists and engineers.

Blackam, E.D., 'The physics of the piano', *Scientific American*, December 1965.
An interesting article describing how the piano works.

Coulson, C.A. and Jeffrey, A., *Waves*, 2nd edn, Longman, 1977.
This is an excellent undergraduate text on the mathematics of wave motion.

Crapper, G.D., *Introduction to Water Waves*, Ellis Horwood, 1984.
A good introduction to the mathematics of the phenomena of waves on water.

Dobbs, E.R., *Electromagnetic Waves*, Routledge and Kegan Paul, 1985.
This text gives an introduction to the mathematics of electromagnetic waves suitable for undergraduate study.

Jacobs, F., 'Using an OHP to demonstrate wave motion', *Physics Education*, Vol. 20, 1985.
An interesting description of how an overhead projector may be used to demonstrate travelling and standing waves and the difference between phase velocity and group velocity.

Pain, H.J., *The Physics of Vibrations and Waves*, 3rd edn, Wiley, 1986.
A very good book written for physics undergraduates and as such puts as much emphasis on the underlying science as on the mathematics.

Ross, D., *Energy from the Waves*, 2nd edn, Pergamon 1981.
A useful introduction to wave energy.

Rossing, T.D., *The Science of Sound*, 2nd edn, Addison-Wesley, 1990.
As well as being a very good text on the science of acoustics, this book provides a good introduction to the general principles of vibrations and waves.

Taylor, C.A., *The Physics of Musical Sounds*, English Universities Press, 1965.

A beautiful book bringing together physics and music. Unfortunately, at the time of writing it is out of print, but most of the material in the book will appear in the following publication: *Experimental Music, the Science of Tones and Tunes*, The Institute of Physics Publishing, 1992.

INDEX